Applied
Mathematical
Physics with
PROGRAMMABLE
POCKET
CALCULATORS

ROBERT M. EISBERG
Professor of Physics
University of California, Santa Barbara

Applied Mathematical Physics with PROGRAMMABLE POCKET CALCULATORS

McGRAW-HILL BOOK COMPANY

New York   St. Louis   San Francisco   Auckland
Bogotá   Düsseldorf   Johannesburg
London   Madrid   Mexico   Montreal
New Delhi   Panama   Paris   São Paulo
Singapore   Sydney   Tokyo   Toronto

APPLIED MATHEMATICAL PHYSICS WITH
PROGRAMMABLE POCKET CALCULATORS

2 3 4 5 6 7 8 9 0 DODO 7 8 3 2 1 0 9 8 7

This book was set in Times Roman by Textbook Services, Inc.
The editors were C. Robert Zappa and Michael Gardner;
the cover was designed by Nicholas Krenitsky;
the production supervisor was John F. Harte.
The drawings were done by J & R Services, Inc.
R. R. Donnelley & Sons Company was printer and binder.

Library of Congress Cataloging in Publication Data

Eisberg, Robert Martin.
    Applied mathematical physics with programmable pocket calculators.

    Includes index.
      1. Differential equations—Numerical solutions.
2. Calculating-machines. 3. Mathematical physics.
I. Title.
QC20.7.D5E35   530.1'5'17   76-23107
ISBN 0-07-019109-3

TO JOANN

CONTENTS

PREFACE

In writing about elementary physics at the level of a college course for engineers and scientists, a textbook author is constantly faced with a serious difficulty. Although it can be assumed that the intended reader knows or is learning some calculus, it cannot be assumed that the reader knows anything about differential equations. But, starting with Newton's law of motion, many of the most important topics of physics lead to a differential equation. If the equation is particularly simple, the author may carefully develop an analytical procedure for solving it and hope that the reader will be able to understand. Frequently, no attempt will be made to treat the topic mathematically because the analytical solution of the equation is just too complicated to be followed by someone who has not yet formally studied differential equations. Then the author must quote results of importance to physics with little or no mathematical justification, to the frustration of both the author and the reader.

A way around this difficulty is to solve the differential equation by a numerical procedure. Compared to analytical methods, numerical methods have several significant advantages: (1) they are so simple in concept that anyone who understands the definition of a derivative will have no difficulty in immediately understanding how they are used to solve differential equations (there is nothing more to them than a succession of multiplications and additions); (2) they are always applied in essentially the same way to all differential equations (whereas the analytical procedure for solving one equation may be drastically different from that which must be used to solve another); and (3) they are almost always successful (even though many important differential equations cannot be solved by any analytical procedure).

A principal disadvantage of any numerical procedure for solving differential equations is that it involves very many repetitious (albeit trivial) calculations, and so is practical only with the assistance of some sort of computing facility.

In the past few years since computers have become widely available on college campuses, several authors have written material showing how they can be used to teach mathematical physics, starting with the elementary course. But, with a few notable exceptions, conventional computers seem to have had little impact on elementary students. This may be because even the smallest conventional computer intimidates many students. To use it the student must spend more time than may be available learning a special programming language. If the student does not type, there may be difficulty in operating the terminal. The computer may not be available at a convenient location and time. And the student is embarrassed about mistakes because they are being made in public on a machine whose costs are not low.

I believe that the advent of programmable pocket calculators provides an almost perfect solution to the difficulty. These calculators, which are actually miniature computers, are very easy to program and operate, inexpensive enough to be widely available, yet powerful enough to solve any of the ordinary differential equations that would be of interest in elementary physics. This includes almost any linear or nonlinear second-order equation, and even sets of two coupled second-order equations. Thus they can be used for topics such as: fall with friction; harmonic and anharmonic oscillations; damped, driven oscillations; coupled oscillations; motion in an attractive or repulsive central force proportional to any power of the distance; and Schroedinger's equation for oscillator and finite square-well potentials. In fact, these are the topics I have chosen to treat in this short book.

In addition, the first two chapters present numerical procedures for differentiating and integrating. The motivation is primarily pedagogical—to start the development of numerical methods and their implementation on programmable pocket calculators with topics that the reader is either familiar with or currently studying by analytical methods. (There is also a practical motivation for numerical integration.) In the next to last chapter, a diversion from the main stream of the book is taken to show how programmable pocket calculators can be used for statistical simulation and analysis by treating a topic involving entropy and the natural direction of flow of time.

The book is written so that it will be particularly easy to use with the HP-25 calculator (produced by Hewlett-Packard, Cupertino, California) or with the SR-56 calculator (produced by Texas Instruments, Dallas, Texas). All of the programs appear in two versions—one suitable for each. At the time of writing, these are the least expensive calculators which have the necessary capability of conditional branching. The programs are also useable, with at most minor modifications, on more sophisticated currently available pocket or desk-top programmable calculators. If a calculator uses algebraic notation, then its programs will follow very closely the SR-56 versions; if it uses parenthesis-free notation, they will be very much like the HP-25 versions. It is reasonable to expect that inexpensive programmable calculators produced in the future will be more versatile and faster, so they will certainly be more than capable of running the programs in this book.

The most ideal situation would be one in which each student has a programmable pocket calculator to work with in the same conditions of privacy available when studying a textbook. For many students this has been or can be realized now, and it does not take much of an extrapolation of current trends to predict that it could soon be universal. Until then, it would certainly be possible for any educational institution to purchase a few calculators to be kept in security cradles on a table in an accessible room. Comparing the ratio of their potential impact on a student's ability to learn physics to their price, with the same ratio evaluated for lecture demonstration or student laboratory equipment, it would seem to be a very cost-effective educational investment.

The only significant physics prerequisite for this book, except for the last chapter, is a knowledge of Newton's law of motion. Although many of the differential equations that are solved have important applications in electrical systems, only their applications to mechanical systems are emphasized. I did this in the expectation that the book could be of use to people whose primary interest is in applied mathematics, and under the assumption that mechanical systems would be more familiar to them. I have tried seriously to make the last chapter, concerning Schroedinger's equation, as self-contained as size limitations allow. But a little contact with elementary quantum physics would certainly make that chapter easier to follow.

As for the mathematical prerequisites, I have tried to make them as few as possible. A prior or concurrent course in elementary calculus, while not an absolute requirement, would be desirable. But it definitely is not necessary to have any background in differential equations to profitably study this book. That does not mean that the numerical methods developed in it will be uninteresting or unuseful for readers who have considerable familiarity with the analytical methods for solving differential equations.

The book is written to be used by a person who has had no exposure to a pocket calculator, programmable or nonprogrammable, or to its instruction manual. This attribute was tested by such a person, who worked with the book during its development.

I would like to emphasize that in no way is it my intent to *replace* the standard analytical methods used to treat topics of elementary mathematical physics. What I am trying to do is to take advantage of technological developments which make it possible to *supplement* these methods, and thereby enhance the ability of students to learn the subject. Many students find numerical methods easier to understand than analytical ones because they are simpler and deal with actual numbers. I believe that all students will benefit very much from studying both methods, since understanding each method will help them to understand the other better. And if they continue on in engineering or science they will, of course, repeatedly come back to numerical methods which are so widely used in professional work.

My deepest appreciation goes to my wife, Lila, who went way beyond the call of duty in testing the book by working with it during its development. I would also like to thank Alice Macnow and C. Robert Zappa of McGraw-Hill for their assistance in the publication of this book and, in particular, for their willingness to publish it despite its unusual character.

Robert M. Eisberg

Applied Mathematical Physics with PROGRAMMABLE POCKET CALCULATORS

CHAPTER ONE

NUMERICAL DIFFERENTIATION

1-1. INTRODUCTION

This chapter will introduce you to the application of numerical methods to topics of mathematics that are central to the study of physics. It is likely that you already have studied, or soon will study, many of these same topics in math and physics courses by using the more conventional analytical methods. The intent is not to try to replace the analytical methods by the numerical ones, but instead to use the latter to supplement the former. Although numerical methods have tremendous advantages of great simplicity and universal applicability, the two approaches are really not in competition; analytical methods are better suited to doing certain things and numerical methods to doing others.

Some subjects that are of interest to elementary physics because of their importance must be treated numerically because analytical methods are not universally applicable. But it is more commonly the case that, although successful analytical treatments of an important physical system are available, they are much too complicated to be used at the elementary level. In these cases the great simplicity of the numerical treatments allows them to be used instead. You will see many examples of this in later chapters. Another situation that frequently arises in elementary mathematical physics is one in which a topic can be handled fruitfully by both methods. In these cases your study of each method can assist your study of the other because the insight you gain into one will enhance your insight into the other.

A good example is differentiation, a topic that certainly plays a vital role in elementary mathematics and physics. This chapter contains two programs, both of which make your programmable calculator automatically evaluate the derivative of any function $f(x)$ you might be concerned with, at any value $x = x_1$ that you want. The first program follows directly from the definition of a derivative, and the familiar figure used to illustrate

1

it, that you will find in any elementary calculus or calculus-level physics textbook. It simply carries out the process that you have probably read and heard so much about: Evaluate $f(x)$ at x_1 and also at the nearby point $x_1 + \Delta x$; subtract the former from the latter to calculate Δf, the change in f; divide by Δx to obtain the average rate of change $\Delta f/\Delta x$; then let Δx become smaller and smaller to find the limit as $\Delta x \to 0$ of $\Delta f/\Delta x$. The program does not just talk about this fundamentally important procedure, it actually does it. Using the program will give you concrete, hands-on experience with the concept of $\Delta f/\Delta x$ approaching its limit.

The second program involves a small change in the first and is based on a slight variation in the standard definition of a derivative that you may not have seen. In using it you will find that, for the same $f(x)$ and x_1, the value of $\Delta f/\Delta x$ approaches the same limit as is obtained with the first program. But it approaches the limit much more rapidly. This will introduce you to a consideration of great practical importance that you will see emphasized repeatedly in later chapters, i.e., rapidity of convergence.

These programs, particularly the second one, provide a practical method of evaluating derivatives numerically. If you are now in the process of learning to differentiate by the conventional analytical method, you can use the second program for checking your mastery of the method (and the accuracy of your homework). First, choose a form for $f(x)$ and a value of x_1, find the analytical form of the derivative in the conventional way, and evaluate it at x_1 by using the calculator manually. Then, insert $f(x)$ and x_1 into the program, run it, and watch the results displayed converge to the limit. Finally, compare to see if your work in finding the analytical form for the derivative, and also your evaluation of it at x_1, were correct. Even if you are way beyond this stage in your mathematical education, you will enjoy doing it a few times.

Chapter 2 gives a completely parallel treatment of numerical integration. Its programs can also be used effectively by beginners for self-testing and by experts for amusement. There is, however, more to numerical integration than that. If an elementary function (i.e., algebraic and/or transcendental) has a derivative, it can always be evaluated by analytical methods. But the same is not true for the integral. As you will see, there are important elementary functions for which the integral can be evaluated only by numerical methods such as are provided by the programs, particularly the second one.

A major purpose of this chapter is to acquaint you with the programmable pocket calculator. The chapter is written under the assumption that you have had no prior contact with it or its instruction manual. As a consequence, the pace is very deliberate. You can speed things up by skipping the explanations of its operation and programs to the extent appropriate to your familiarity with the calculator, but be sure to read the other material as it will be referred to subsequently.

1-2. DERIVATIVES BY FULL-INCREMENT METHOD

The *derivative* $df(x)/dx$ of the function $f(x)$ at $x = x_1$ is defined as:

$$\left[\frac{df(x)}{dx}\right]_1 \equiv \text{limit as } \Delta x \text{ approaches } 0 \text{ of } \left[\frac{f(x_1 + \Delta x) - f(x_1)}{\Delta x}\right]$$

The definition is illustrated in Fig. 1-1 and is described in words in the preceding section. The derivative of $f(x)$ at x_1 is nothing more than its rate of change at that point, and is measured by the *slope* of the straight line that is tangent to the curve there. However, if you have never been exposed to this definition before, it would be wise to look at one of the books in Ref. 1. The concept of a derivative will be used throughout almost all of this book, but it will not actually be necessary for you to know how to obtain derivatives analytically. All you will need to know is what a derivative is, and you will certainly learn that in working through this chapter.

Programs carrying out the procedure implied by the definition are listed in Table 1-1 (HP-25) for the Hewlett-Packard HP-25 calculator and others using *parenthesis-free logic,* and, separately, in Table 1-1 (SR-56) for the Texas Instruments SR-56 and others using *algebraic logic*.

The two programs differ in detail, but are exactly the same in what they do. After the appropriate general program is keyed into your calculator, choose $f(x)$ and key into an ample blank place in the program the steps which will start from x and generate the chosen form of $f(x)$ using, if necessary, some of the storage registers not used by the program. Then, in-

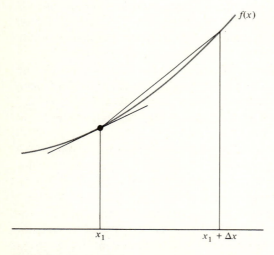

Figure 1-1. Illustration of the full-increment definition of a derivative.

sert whatever values you choose of x_1, and of the first Δx to be used, in the designated storage registers and push the button that starts the calculator. It will evaluate $[f(x_1 + \Delta x) - f(x_1)]/\Delta x$ and stop with the result displayed. Inspect the result, writing it down if you wish, and then push the start button again. The calculator will reduce Δx by a factor of $\frac{1}{2}$, repeat the calculation, and then stop with the new result displayed. This cycle continues every time you restart the calculator. The remainder of this section shows you, by example, how to enter and use the program for your calculator, but does not explain the operations performed by the calculator when it is running the program. Such an explanation is given in the next section.

Example
$f(x) = x^3$, $x_1 = 2$, initial $\Delta x = 1$

For the HP-25
Switch to **PRGM** and clear by keying **f PRGM.** Then key steps 01 through 04 of the program. To do this, simply follow in sequence the key entries listed in the third column of Table 1-1 (HP-25), noting that frequently there are two or even three key entries per step. To be specific, key **RCL 1.** This completes step 01, and the display shows 01   24 01. The first two numbers are the step identification. The second pair identify the key entry **RCL** by specifying its position on the keyboard—**RCL** is in row 2 from the top and column 4 from the left. The third pair of numbers identify the key for the digit 1. The display makes it easy to check your program keying for errors. Next, key **RCL 2** and note that the display shows 02   24 02. The third step is obtained by keying + , which produces the display 03 51. Then key **STO 3,** and note that the display shows 04   23 03. If you are careful in keying, it is usually unnecessary to watch the display while doing it. If you feel you have made a mistake in a step preceding the one currently displayed, each press of **BST** will make the display go back one step. An error in, say, step 02 can be corrected by pressing **BST** once more so that step 01 is displayed, and then correctly keying step 02. Pressing **SST** will allow you to go forward in single steps to where you were initially.

Now you come to steps 05 through 31, in which there is room to use the calculator's built-in functions and the unused storage registers to construct almost any form of $f(x)$ that you would like. For $f(x) = x^3$, this is done by keying **3** for step 05, **f y^x** for step 06, and **GTO 3 2** for step 07. The latter will cause the calculator to bypass the unused steps when a program is being run. Next, press the **SST** key repeatedly until the pair of numbers on the left side of the display read 31, and then continue entering the program by keying **STO 4, RCL 3,** etc. When you finish, switch to **RUN,** and set the program to step 00 by keying **f PRGM.** Load x_1, and the initial Δx, by keying **2 STO 1,** and **1 STO 2.** Now you can run the program by pushing the **R/S** button. To run it again, push **R/S** again, etc.

For the SR-56

Clear the program by keying **2nd CP.** Then prepare the calculator to *learn* the program by keying **LRN.** The display will show 00 00. The first pair of numbers tells you it is ready to receive step 00, and the second pair shows that nothing is currently in that step. Now follow in sequence the key entries listed in the third column of Table 1-1 (SR-56). To be specific, first key **RCL.** The display will change to 01 00, indicating that you should now enter step 01. Do so by keying the digit **1.** Next, key **+.** Any time you feel you have made a mistake, press **2nd bst** as many times as required to review the previously entered steps by means of a code identifying the keys that were entered. If you do it once after having keyed the + for step 02, you should see 02 84 on the display. The second pair of numbers is the code for the location of the + key, which is in row 8 from the top and column 4 from the left. If it is incorrect, reenter the proper key and the error will be corrected. You can backstep farther by continued use of **2nd bst.** To return to where you were, press **SST** to single step forward. Now key through step 06. Then push **2nd bst** and you will see 06 57. The code indicates that the operation in step 06 is **2nd subr;** it is the second function of the key labeled **(** , which is the one shown above the key. The last digit in the code for a second function starts with 6 for the first column on the left. So 57 means row 5 from the top, second function on column 2 from the left. Before entering step 07, press **SST** once to get back to the proper location for that step. Then continue entering the program through step 33.

Starting at step 34, there is room to use the calculator's built-in functions and all the unused storage registers to make a subroutine for constructing almost any form of $f(x)$ you would like. For $f(x) = x^3$, this is done by keying y^x for step 34, **3** for step 35, and = for step 36. Always end the subroutine by keying **2nd rtn.** Now press **LRN** again to get out of the learn mode and press **RST** to set the calculator to step 00. Load x_1, and the initial value of Δx, by keying **2 STO 1,** and **1 STO 2.** You can now run the program by pressing the **R/S.** To run again, press **R/S** again, etc.

This ends the instructions for running the example on the HP-25 or the SR-56.

For either calculator the results, rounded off to three decimal places by keying **f FIX 3** on the HP-25 and **2nd fix 3** on the SR-56, are:

$$\Delta f / \Delta x = 19.000, \ 15.250, \ 13.562, \ 12.766, \ 12.379, \ 12.188, \ 12.094,$$
$$12.047, \ 12.023, \ 12.012, \ 12.006, \ 12.003, \ 12.001, \ 12.001,$$
$$12.000, \ 12.000, \ 12.000$$

The results have converged, although not very rapidly, to the value 12.000. This is, of course, in agreement with the analytical method which predicts a value of 12. If you know how to evaluate derivatives analytically, verify this.   ////

Table 1-1. (HP-25) Derivative by full-increment method

Register	0	1	2	3	4	5	6	7
Register Contents:	—	x_1	Δx	x_i	$f(x_i)$	$\Delta f/\Delta x$	—	—
		preloaded						

Program

Step	Code	Key Entry	X	Y	Z	T	Comments
00							
01	24 01	RCL 1	x_1				Start loop
02	24 02	RCL 2	Δx				
03	51	+	$x_1 + \Delta x$				
04	23 03	STO 3	x_i	x_1			
05							$x_i = x_1 + \Delta x$, initially Steps 05 through 31, plus registers 0, 6, and 7 available for constructing $f(x_i)$. If fewer steps used, end with **GTO 32**
.							
.							
.							
.							
31			$f(x_i)$				
32	23 04	STO 4	$f(x_i)$				$x_i = x_1 + \Delta x$, or x_1
33	24 03	RCL 3	x_i	x_i			Stored until x_i identified
34	24 01	RCL 1	x_1	x_i			
35	14 51	f x ≥ y	x_1				Test x_i to identify content of register 4

6

Step	Code			Keystroke	X		Comment
36		13	41	GTO 41		x_i	
37		24	04	RCL 4	x_1		
38		23	05	STO 5	$f(x_1 + \Delta x)$		5 now $f(x_1 + \Delta x)$
39		24	01	RCL 1	$f(x_1 + \Delta x)$		To evaluate $f(x_1)$
40		13	04	GTO 04	x_1		
41		24	04	RCL 4	x_1		
42	23	41	05	STO − 5	$f(x_1)$		
43		24	02	RCL 2	$f(x_1)$		5 now Δf
44	23	71	05	STO ÷ 5	Δx		
45		24	05	RCL 5	Δx		5 now $\Delta f/\Delta x$
46			74	R/S	$\Delta f/\Delta x$		
47			02	2	$\Delta f/\Delta x$		
48	23	71	02	STO ÷ 2	2		Δx reduced
49		13	01	GTO 01	2		To new loop

Table 1-1. (SR-56) Derivative by full-increment method

Register Contents:

0	1	2	3	4	5	6	7	8	9
—	x_1	Δx	$\Delta f/\Delta x$	—	—	—	—	—	—
	preloaded								

Program

Step	Code	Key Entry	Comments
00	34	RCL	Start loop
01	01	1	x_1
02	84	+	
03	34	RCL	Δx
04	02	2	$x_1 + \Delta x$
05	94	=	To evaluate $f(x_1 + \Delta x)$
06	57	2nd subr	
07	03	3	
08	04	4	
09	33	STO	3 now $f(x_1 + \Delta x)$
10	03	3	
11	34	RCL	x_1
12	01	1	To evaluate $f(x_1)$
13	57	2nd subr	
14	03	3	
15	04	4	
16	12	INV	
17	35	SUM	3 now Δf
18	03	3	
19	34	RCL	

Step	Code	Key Entry	Comments
20	02	2	Δx
21	12	INV	
22	30	2nd PROD	3 now $\Delta f/\Delta x$
23	03	3	
24	34	RCL	$\Delta f/\Delta x$
25	03	3	
26	41	R/S	2
27	02	2	
28	12	INV	
29	30	2nd PROD	Δx reduced
30	02	2	
31	22	GTO	
32	00	0	To new loop
33	00	0	
34	.	.	Steps 34 through 99, plus
.			registers 0, 4, 5, 6, 7, 8, and 9
.			available for constructing $f(x)$
.			from x, which is in display
			register at 34. End with
			2nd rtn

8

The sequence of results was terminated when, as seen with three decimal places, the repeated values 12.000 clearly indicated that convergence had been achieved. But what happens if you nevertheless continue to cycle the calculator? Try it. You will find that before long the results begin to fluctuate about the value of 12.000. If continued, the fluctuations grow until the results become meaningless. This behavior has nothing to do with derivatives, except that their definition involves the difference between two numbers. It actually arises from roundoff error in the calculator, and if you think about what happens for a minute you should be able to devise an explanation if you understand what a derivative is and what the calculator is doing. In practice the fluctuations are not a significant limitation to the utility of the numerical method of obtaining derivatives, since the value of the limit can be seen with adequate accuracy before the fluctuations set in. This is particularly true for the method described in Sec. 1-4, which converges more rapidly.

If you want to try different values of x_1 and/or the initial Δx, still using $f(x) = x^3$, do the following:

For the HP-25
Key **f PRGM** to reset to step 00; key x_1 **STO 1**, initial Δx **STO 2**; then push **R/S.**

For the SR-56
Key **RST** to reset to step 00; key x_1 **STO 1**, initial Δx **STO 2**; then push **R/S.**

To change $f(x)$ to a different form, proceed with the following instructions.

For the HP-25
While switched to **RUN,** key **GTO 0 4,** switch to **PRGM,** then key the steps required to construct $f(x)$ from x, assuming the result of step 04 is to display x. Steps 05 through 31 are available, and also registers 0, 6, and 7. If you use fewer steps, end by keying **GTO 3 2.** Then switch to **RUN,** and reset to step 00 by keying **f PRGM.** Next key x_1 **STO 1**, Δx **STO 2.** Start by pushing **R/S.**

To be specific, if you want to change to $f(x) = x^n$, where n is any integer from 0 to 9, just key in n instead of **3** in step 05. If n requires more than one key entry, the remaining steps used to generate $f(x)$ will have to be keyed in again because more steps will be used to enter n.

As another example, for $f(x) = 2x + 3\cos x$, key in the following: Step 05 is **STO 0**, 06 is **2**, 07 is ×, 08 is **RCL 0**, 09 is **f cos**, 10 is **3**, 11 is ×, 12 is +, 13 is **GTO 3 2**. After switching to **RUN,** and keying **f PRGM,** be sure to key **g RAD** to make the calculator operate with radians.

To see a listing of the key entries used to generate a more complicated function, see Sec. 2-4.

For the SR-56

While still in the program execution mode, key **GTO 3 4**, key **LRN**, then key the steps required to construct $f(x)$ from x, assuming the result of step 33 is to put x in the display register. The steps through 98 are available, and also registers 0 and 4 through 9. The last step of the subroutine must be **2nd rtn.** Then key **LRN**, and reset to step 00 by keying **RST**. Next, key x_1 **STO** 1, Δx **STO** 2. Start by pushing **R/S**.

To be specific, if you want to change to $f(x) = x^n$, where n is any integer from 0 to 9, just key n instead of **3** in step 35. If n requires more than one key entry, the remaining steps used to generate $f(x)$ will have to be keyed again because more steps will be used to enter n.

As another example, for $f(x) = 2x + 3\cos x$, key in the following: Step 34 is **STO**, 35 is **0**, 36 is ×, 37 is **2**, 38 is +, 39 is **3**, 40 is ×, 41 is **RCL**, 42 is **0**, 43 is **cos**, 44 is =, 45 is **2nd rtn.** After pushing **LRN**, and then **RST**, be sure to key **2nd RAD** to make the calculator operate with radians.

To see a listing of the key entries used to generate a more complicated function, see Sec. 2-4.

1-3. EXPLANATION OF PROGRAM OPERATION

For both versions of the program, the columns to the right of the column listing the key entries show what is happening when the program is being run. You can ignore them now and throughout this book if you wish. The major part of the Exercises at the end of each chapter does not require a knowledge on your part of exactly how the programs operate and, as you have just seen, the programs can be keyed into the calculator by simply following the instructions in the third column. However, the operation of the programs is easy enough to understand for it to be likely that in the process of entering and using a few of them, you will learn what they are doing without formally studying programming, provided you are given a few clues at the beginning. If this interests you, read one of the following explanations.

Although the program explained is the first one in this book, for the HP-25 it is almost as sophisticated as any of them. So if you can understand its operation with a little effort, you surely will be able to understand how all of them operate. Then you will be able to obtain the even greater satisfaction of writing your own variations on a program, or your own original programs. Help can be found in the calculator instruction manual, of course, and also in Sec. 3-4 where there is another detailed explanation of the operation of a program. For the SR-56, the first program does not require the use of a test routine that is needed in the two programs of Chap. 2, and in Exercise 1-6. So if you are using that calculator, it might be appropriate to read the pertinent parts of Sec. 3-4 after you read this section.

For the HP-25

In step 01 the contents of register 1, which will be x_1, are recalled and placed in the lowest, or X register, of four operating registers called the *stack*. The entry in the column labeled X shows what is in that register after the operation for that step has been carried out, and similarly for the other columns and their associated stack registers. In step 02, register 2 is recalled and its contents Δx are placed in X. One property of the stack is that when a number is inserted in its X register, any number already there will be displaced up to the next higher, or Y register. And if something was already in Y (not so here), it would be displaced up to Z, etc. In step 03 the contents of X are added to the contents of Y. The results of this operation, or of any other arithmetic operation carried out on the X and Y registers, are put in the X register. Furthermore, if there was something in the Z register initially (not so here), it would be displaced down and end up in the Y register, and similarly for T. Thus step 03 produces $x_1 + \Delta x$ in X. In step 04 this is stored in register 3, so that it will initially contain $x_i = x_1 + \Delta x$.

Steps 05 through 31 are chosen by the user to construct $f(x_i)$ from the x_i present in X at step 04. In the example, $f(x) = x^3$; step 05 is **3**; step 06 is **f** y^x; and step 07 is **GTO 3 2**. The first of these steps puts the digit **3** in X, displacing x_i to Y. The next step takes the contents of Y to the power specified by the contents of X, thus producing x_i^3. (The prefix key **f** is used to specify which of the three labels, y^x **3 ABS,** associated with the same key is the one intended.) Just as in addition, the results of this arithmetic operation are put into X. Step 07 makes the calculator skip directly to step 32 where the contents of X, namely $f(x_i)$, are stored in register 4.

Now it gets interesting. Step 33 puts the contents of 3, which is the value of x_i, into X. [$f(x_i)$ is also displaced to Y, but it will no longer be listed since when it is subsequently used it will be obtained not from the stack but by recalling 4.] In step 34, x_1 is put in X, thereby displacing x_i to Y, by recalling register 1. In the next step a test is carried out comparing the contents of the X and Y registers for the purpose of identifying the value of x_i. The reason why it is necessary is that the routine for generating $f(x_i)$ from x_i must be used twice—the first time with $x_i = x_1 + \Delta x$, and the second time with $x_i = x_1$. At this stage of the explanation, you are at the point where $x_i = x_1 + \Delta x$, so the contents of Y is a larger number than the contents of X. The test in step 35 asks the question: Is $x \geq y$, where x and y mean the contents of the X and Y registers? At this stage the answer will be *no*. If the answer to this or any of the other tests available on the calculator is *no*, then in executing a program it will automatically skip the next step. Thus this time it will go to step 37. In step 37, register 4 is recalled, putting $f(x_1 + \Delta x)$ into X. Note that the value of x_i has been specified since the test identifies what it is. Step 38 stores $f(x_1 + \Delta x)$ in register 5, and step 39 recalls register 1 to put x_1 in X.

Step 40 makes the calculator return to step 04, with x_1 in X. So step 04 causes x_1 to be stored in 3 this time. Then things proceed as before through step 34, except that now $x_i = x_1$ so that the contents of register 4 are $f(x_1)$. As a consequence, when the test of step 35 is made the answer to the question will be *yes*. If the answer to any of the test questions is *yes*, then the calculator does not skip a step. Thus it will go to step 36 which, in turn, will cause it to go to step 41. In step 41 it recalls the contents of register 4, thereby placing $f(x_1)$ in X. Step 42 causes the number in X to be subtracted from the number in register 5; thus 5 will now contain $f(x_1 + \Delta x) - f(x_1) = \Delta f$. Steps 43 and 44 obtain Δx from register 2 and then divide it into the contents of register 5. Now register 5 will contain $\Delta f/\Delta x$. To make this result visible to you, step 45 recalls 5 to the X register and step 46 stops the calculator. The display always shows what is in X, but it can be seen, of course, only when the calculator is stopped or at least is pausing.

When you push **R/S** to actuate the calculator again, it will go to steps 47 and 48 that cause it to reduce by a factor of $\frac{1}{2}$ the value of Δx stored in register 2. Then it goes to step 49, which causes it to go to step 01 where it immediatley begins to execute a new loop of the calculation with the reduced Δx.

For the SR-56
In steps 00 and 01 the contents of register 1, which will be x_1, are recalled to a register that is capable of being displayed. The x_1 in the Comments column shows the contents of that *display register* after the operation for step 01 has been carried out. Comments are also used on occasion to describe what is happening in words, e.g., the "start loop" listed for step 00. Step 02 prepares to make an addition. In the process, x_1 is removed from the display register and stored elsewhere pending the completion of the addition. Steps 03 and 04 recall the contents of register 2, Δx, to the display register. The addition is completed in step 05 and the result, $x_1 + \Delta x$, ends up in the display register. All of the other arithmetic operations performed on two numbers are handled in essentially the same way as was this addition. At the $=$ instruction the calculator carries out all pending operations, first performing multiplication and division and then performing addition and subtraction. (If an operation involving one number, e.g., taking the sine of the number, is also pending, it will be carried out before the two number operations.) Steps 06, 07, and 08 transfer execution of the program to the subroutine for calculating $f(x_1 + \Delta x)$ from $x_1 + \Delta x$, which starts in step 34.

Steps 34, 35, 36, . . . , plus any of the registers not employed in the main program, are available to the user for constructing the function from its argument, which will be in the display register upon arrival at step 34. In the example, $f(x) = x^3$; step 34 is y^x, step 35 is **3,** and step 36 is $=$. The first of these prepares the calculator to take an as yet unspecified power of $x_1 + \Delta x$; the second specifies the power to be 3; and the third carries out the operation, leaving $f(x_1 + \Delta x) = (x_1 + \Delta x)^3$ in the display register. Step 37 is the mandatory **2nd rtn,** which automatically makes execution return to the step immediately following the last one executed in the main program. In this case the calculator will go back to step 09, carrying $f(x_1 + \Delta x)$ in the display register.

Steps 09 and 10 store that quantity in register 3. Steps 11 and 12 recall register 1, which is x_1, and put it in the display register. Steps 13, 14, and 15 carry it back in the display register to the same subroutine. Now the subroutine calculates $f(x_1)$ and returns with that quantity in the display register to the step after the one that led to the subroutine this time. Thus $f(x_1)$ is in the register for step 16. That step instructs the calculator to perform on $f(x_1)$ the inverse of the operation called for in steps 17 and 18. The net effect is to subtract $f(x_1)$ from the contents of register 3, leaving that register with contents Δf. Steps 19 and 20 recall Δx from register 2 to the display register, where steps 21, 22, and 23 then divide it into the Δf in register 3. As a result, 3 contains $\Delta f / \Delta x$ at the end of step 23. Steps 24 and 25 recall 3 to the display register. Step 26 halts the program and unblanks the display register so that $\Delta f / \Delta x$ can be seen.

After an **R/S** is keyed in manually to restart, step 27 puts the digit 2 in the display register, and then steps 28, 29, and 30 divide it into the contents of register 3. Steps 31, 32, and 33 take the program back to step 00, where it immediately proceeds to start a new loop. This loop goes exactly as the first one, except that it uses the new reduced value of Δx.

1-4. DERIVATIVES BY HALF-INCREMENT METHOD

A minor alteration in the definition of a derivative will make a major improvement in the rapidity of convergence of the sequence of numbers produced in numerical differentiation. The altered definition is

$$\left[\frac{df(x)}{dx} \right]_1 \equiv \text{limit as } \Delta x \text{ approaches 0 of } \left[\frac{f(x_1 + \Delta x/2) - f(x_1 - \Delta x/2)}{\Delta x} \right]$$

and is illustrated in Fig. 1-2. In this so-called *half-increment* definition, the derivative involves evaluating $f(x)$ at two points symmetrically disposed about the point x_1 where the value of the derivative is required. The definition is completely equivalent to the standard definition, which may be called the *full-increment* definition, in that both lead to the same limit as $\Delta x \to 0$. And neither definition is better or worse conceptually. But the half-increment definition produces a much superior numerical method, from a practical point of view, because the results it yields converge to their limit much more rapidly.

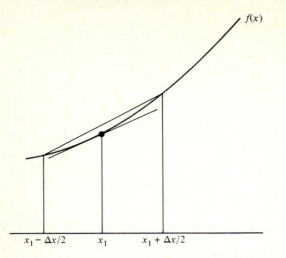

Figure 1-2. Illustration of the half-increment definition of a derivative.

You can see why by inspecting Figs. 1-1 and 1-2. In both, the slope of the straight line, tangent to the $f(x)$ curve at x_1, is a measure of the derivative at that point. The slope of the straight line connecting the value of $f(x)$ at the extreme ends of the interval of total width Δx is the approximation to the derivative generated by the definition for the value of Δx shown in the figure. Clearly, for a given Δx, the altered definition gives a much better approximation. Hence the more rapid convergence.

The alterations in the programs for each of the two calculators are equally minor, as inspecting Tables 1-2 will show you. They are used exactly as before and, for the same example, yield the following results.

Example

$f(x) = x^3$, $x_1 = 2$, initial $\Delta x = 1$

[entered and run in the HP-25 or SR-56 exactly as before, except for the difference in the steps available for constructing $f(x)$]. The results to three decimal places are:

$$\Delta f / \Delta x = 12.250, \ 12.063 \text{ or } 12.062, \ 12.016, \ 12.004, \ 12.001,$$
$$12,000, \ 12,000, \ 12.000$$

(12.063 will be obtained on the HP-25 and 12.062 on the SR-56; the variation arises from the difference in the way the calculators round off to three decimal places.) ////

If you compare these with the results obtained for the same $f(x)$, x_1, and initial Δx using the full-increment method, you will see why the half-increment method is the preferred one to use. And you will see in the next

Table 1-2. (SR-56) Derivative by half-increment method

Register Contents:

0	1	2	3	4	5	6	7	8	9
—	x_1	Δx	$\Delta f/\Delta x$	—	—	—	—	—	—
	preloaded								

Program

Step	Code	Key Entry	Comments
00	34	RCL	Start loop
01	01	1	x_1
02	84	+	
03	34	RCL	
04	02	2	Δx
05	54	÷	
06	02	2	
07	94	=	$x_1 + \Delta x/2$
08	57	2nd subr	To evaluate $f(x_1 + \Delta x/2)$
09	04	4	
10	02	2	
11	33	STO	
12	03	3	3 now $f(x_1 + \Delta x/2)$
13	34	RCL	
14	01	1	x_1
15	74	−	
16	34	RCL	
17	02	2	Δx
18	54	÷	
19	02	2	
20	94	=	$x_1 - \Delta x/2$
21	57	2nd subr	To evaluate $f(x_1 - \Delta x/2)$
22	04	4	
23	02	2	

Step	Code	Key Entry	Comments
24	12	INV	
25	35	SUM	
26	03	3	3 now Δf
27	34	RCL	
28	02	2	Δx
29	12	INV	
30	30	2nd PROD	
31	03	3	3 now $\Delta f/\Delta x$
32	34	RCL	
33	03	3	$\Delta f/\Delta x$
34	41	R/S	
35	02	2	2
36	12	INV	
37	30	2nd PROD	
38	02	2	
39	22	GTO	
40	00	0	Δx reduced
41	00	0	To new loop
42	·		Steps 42 through 99, plus registers 0, 4, 5, 6, 7, 8, and 9 available for constructing $f(x)$ from x, which is in display register at 42. End with **2nd rtn**
·	·		
·	·		

Table 1-2. (HP-25) Derivative by half-increment method

Register:	0	1	2	3	4	5	6	7
Contents:	—	x_1	Δx	x_i	$f(x_i)$	$\Delta f/\Delta x$	—	—
		preloaded						

Program

Step	Code	Key Entry	X	Y	Z	T	Comments
00							
01	24 01	RCL 1	x_1				Start loop
02	24 02	RCL 2	Δx	x_1			
03	02	2	2	Δx	x_1		
04	71	÷	$\Delta x/2$	x_1			
05	51	+	$x_1 + \Delta x/2$				
06	23 03	STO 3	x_i				$x_i = x_1 + \Delta x/2$, initially
⋮							Steps 07 through 27, plus registers 0, 6 and 7, available for constructing $f(x_i)$. If fewer steps used, end with **GTO 28**
27			$f(x_i)$				
28	23 04	STO 4	$f(x_i)$				$x_i = x_1 + \Delta x/2$, or $x_1 - \Delta x/2$
29	24 03	RCL 3	x_i				Stored until x_i identified
30	24 01	RCL 1	x_1	x_i			Test x_i to identify content of register 4
31	14 51	f $x \geq y$	x_1	x_i			

Step	Code	Key	Register/Display	Comment
32	13 41	GTO 41	x_i	
33	24 04	RCL 4	$x_1 + \Delta x/2$	
34	23 05	STO 5	$f(x_1 + \Delta x/2)$	5 now $f(x_1 + \Delta x/2)$
35	24 01	RCL 1	x_1	
36	24 02	RCL 2	Δx	
37	02	2	2	x_1
38	71	÷	$\Delta x/2$	x_1
39	41	−	$x_1 - \Delta x/2$	
40	13 06	GTO 06	$x_1 - \Delta x/2$	To evaluate $f(x_1 - \Delta x/2)$
41	24 04	RCL 4	$f(x_1 - \Delta x/2)$	
42	23 41 05	STO − 5	Δx	5 now Δf
43	24 05	RCL 5	Δx	
44	23 71 05	STO ÷ 5	$\Delta f/\Delta x$	5 now $\Delta f/\Delta x$
45	24 05	RCL 5	$\Delta f/\Delta x$	
46	74	R/S		
47	02	2	2	
48	23 71 02	STO ÷ 2	2	Δx reduced
49	13 01	GTO 01	2	To new loop

chapters that this conclusion applies even more strongly to numerical integration and to the numerical solution of differential equations.

EXERCISES

1-1. Run the full-increment and half-increment programs to evaluate the derivative of $f(x) = x^3$ at several values of x_1. For each of these, compare the speeds of convergence of the two programs.

1-2. Try other functions such as power laws, sinusoidals, and exponentials in the two programs. For what functions is the discrepancy in speed of convergence between the two programs most striking? Why?

1-3. Use the half-increment program to test your knowledge of the analytical expression for the derivative of a variety of functions and combinations of functions. For each case run the program at some value of x_1 and then compare the results with what you get by using the calculator to evaluate the analytical expression at that value of x_1.

1-4. Using the half-increment program to evaluate the derivative of some function at a number of uniformly spaced values of x_1, make a plot of its derivative versus x_1. Then use the calculator to evaluate the function at the same values of x_1 and plot them. Compare the plots and comment on them. Interesting functions to consider are $\sin x$ and $\cos x$.

1-5. Study the details of the half-increment program and explain precisely what happens in each step.

1-6. Write a modification to the half-increment program that will cause it to stop and display the final result when Δx becomes less than some preloaded value of δ. How should δ be chosen? Is it better to stop when Δf becomes sufficiently small or, perhaps, when the change in $\Delta f / \Delta x$ from one loop to the next becomes sufficiently small? Write program modifications to try these.

1-7. Write a half-increment program to differentiate a vector, the vector being the velocity of a particle moving uniformly around a circle, and the derivative taken with respect to time. Compare the results with the analytical expression for the centripetal acceleration of the particle.

REFERENCES

1. Bueche, Frederick J.: *Introduction to Physics for Scientists and Engineers,* 2d ed., McGraw-Hill Book Company, New York, 1975, p. 34.

    Halliday, David, and Robert Resnick: *Fundamentals of Physics,* John Wiley & Sons, Inc., New York, 1974, p. 27.

    Sears, Francis W., and Mark W. Zemansky: *University Physics,* 4th ed., Addison-Wesley Publishing Company, Inc., Reading, Mass., 1970, p.42.

    Stein, Sherman K.: *Calculus and Analytic Geometry,* McGraw-Hill Book Company, New York, 1973, pp. 38–43.

TWO

NUMERICAL INTEGRATION

2-1. INTRODUCTION

In this chapter you will again start with a program which is based directly on a common textbook definition—in this case, of an integral. Then you will again find that a slight variation in the definition, and a correspondingly small change in the program, produces a very superior numerical method. The second method of evaluating definite integrals, called the half-increment method, is superior because it converges to a result of useable accuracy very much more rapidly than does the first, or full-increment method. This advantage is much more important than it is in the case of numerical differentiation. Even the slower differentiation program is quite fast, while the slower integration is really too slow to be practical.

Are *any* of these programs of practical value? The principal justification for numerical differentiation is pedagogical—it can help you master the techniques and concepts involved in differentiation by analytical methods. There is the same justification for numerical integration. But there is an additional justification for studying numerical integration which is actually much more significant, i.e., there are many important elementary functions which cannot be integrated analytically. For these the *only* available methods of integration are numerical ones.

An example which you will see worked out explicitly at the end of this chapter is the integral over a certain wavelength range of the function which specifies the spectrum of radiation emitted by the sun. When a solar physicist (or an engineer working with the design of a solar-heating plant) needs to know how much power is emitted in a particular range of the spectrum, there is no choice but to employ numerical methods to evaluate the integral.

2-2. DEFINITE INTEGRALS BY FULL-INCREMENT METHOD

Inspect Fig. 2-1. Have you seen a figure before that looks something like it? If so, then you probably are familiar with the so-called *definite integral* $\int_{x_2}^{x_1} f(x)\,dx$ of the function $f(x)$ from $x = x_1$ to $x = x_2$. The quantity is equal to the area under the curve of $f(x)$ between these two points, and is defined in terms of the area under the rectangles as

$$\int_{x_1}^{x_2} f(x)\,dx \equiv \text{limit as } \Delta x \text{ approaches } 0 \text{ of } \left[\sum_{x_i} f(x_i)\,\Delta x \right]$$

where $x_i = x_1,\ x_1 + \Delta x,\ x_1 + 2\Delta x,\ \ldots,\ x_2 - \Delta x$. If the figure is familiar, but the mathematical expression of the definition is not, you may be able to understand the latter by comparing it with the former if you have the hint that $\sum_{x_i}$ means "summation over the indicated range of values of x_i." But if all of this is new, you should first look at one of the books of Ref. 1 at the end of this chapter.

Programs shown in Tables 2-1 (SR-56) and (HP-25) carry out the operations indicated by the definition. After the general program appropriate to your calculator is keyed in, choose $f(x)$ and key into the empty region the steps which will generate $f(x)$ from x. Use any of the available storage registers, if required. Then preload two designated storage registers with the values selected for x_1 and x_2, and push the start button. The calculator will then evaluate $\sum_{x_i} f(x_i)\,\Delta x$, initially taking $\Delta x = x_2 - x_1$, and stop with the results displayed. On the next push of the button, it will reduce Δx by a factor of $\frac{1}{2}$, repeat the calculation, and again stop with the results displayed.

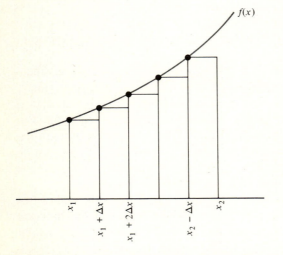

Figure 2-1. Illustration of the full-increment definition of a definite integral.

Example

$f(x) = x^2$, $x_1 = 1$, $x_2 = 2$

For the SR-56
Clear by keying **2nd CP**, key **LRN**, then enter the program by keying steps
00 through 50. Step 51 is x^2 and step 52 is **2nd rtn.** Key **LRN** to return to
the execute mode, and key **RST** to set to step 00. Load x_1 and x_2 by keying
1 STO 1 and **2 STO 2.** Then push the **R/S** button. To rerun, push **R/S**
again.

For the HP-25
Switch to **PRGM**, clear by keying **f PRGM**, then key steps 01 through 11 of
the program. Step 12 is **g** x^2 and step 13 is **GTO 3 7**. Next press **SST** until
you reach step 36, and then key in the remainder of the program. Switch
to **RUN**, and set to step 00 by keying **f PRGM**. Load x_1 and x_2 by keying **1**
STO 1 and **2 STO 2.** Then push the **R/S** button. To rerun, push **R/S** again.

For either calculator the results, rounded off to two decimal places by
keying **2nd fix 2** on the SR-56 and **f FIX 2** on the HP-25, are:

$$\sum_{x_i} f(x_i) \Delta x = 1.00,\ 1.63,\ 1.97,\ 2.15,\ 2.24,\ 2.29,\ 2.31,\ 2.32,\ 2.33,\ 2.33$$

If you are willing to wait, this numerical integration method does converge to the value $\frac{7}{3}$ predicted by the analytical method. But each result
unavoidably takes twice as much time to obtain as the previous result
because it comes from computing and summing twice as many terms. And
since the results obtained converge to their limit only very gradually, the
net effect is that integration by this method is very time-consuming. ////

If, nevertheless, you want to try different values of x_1 and/or x_2, or to
try different forms for $f(x)$, consult the instructions for making these
changes in running the differentiation programs of the previous chapter.
Except for the obvious modifications required to account for the facts that
x_2 instead of Δx is to be loaded in a register, and that the steps available
for constructing $f(x)$ are different, these instructions apply exactly here.
It may be worthwhile using this *full-increment* method to make a
numerical integration of a function that you can also integrate analytically, for checking purposes. But do not do too much with it before look-
ing at the half-increment method presented in the next section.

2-3. DEFINITE INTEGRALS BY HALF-INCREMENT METHOD

The *half-increment* method uses the slightly altered definition of a definite
integral illustrated in Fig. 2-2. Comparison with Fig. 2-1 makes it clear
why the half-increment definition provides the basis for a superior

Table 2-1. (SR-56) Definite integral by full-increment method

Register Contents:

0	1	2	3	4	5	6	7	8	9
—	x_1	x_2	Δx	x_i	Σ	—	—	—	—
	___ preloaded ___								

Program

Step	Code	Key Entry	Comments
00	34	RCL	
01	02	2	x_2
02	32	x ⋛ t	Test register loaded with x_2
03	34	RCL	
04	02	2	x_2
05	74	−	
06	34	RCL	
07	01	1	x_1
08	33	STO	
09	04	4	4 set to x_1
10	94	=	$\Delta x = x_2 - x_1$, initially
11	33	STO	
12	03	3	

Step	Code	Key Entry	Comments
28	35	SUM	
29	05	5	$f(x_i)\,\Delta x$ added to Σ
30	34	RCL	
31	03	3	Δx
32	35	SUM	
33	04	4	x_i incremented
34	22	GTO	
35	01	1	
36	06	6	To evaluate next $f(x_i)\,\Delta x$
37	34	RCL	
38	05	5	Σ
39	41	R/S	
40	02	2	2

Step	Code	Key	Comment
13	00	0	0
14	23	**STO**	
15	05	5	5 zeroed
16	34	**RCL**	
17	04	4	
18	47	**2nd x ≥ t**	
19	03	3	
20	07	7	Tests for x_i at end of interval
21	57	**2nd subr**	To evaluate $f(x_i)$
22	05	5	
23	01	1	
24	64	×	
25	34	**RCL**	
26	03	3	x_i
27	94	=	Tests for x_i at end of interval

Step	Code	Key	Comment
41	12	**INV**	
42	30	**2nd PROD**	
43	03	3	Δx reduced
44	34	**RCL**	
45	01	1	x_1
46	33	**STO**	
47	04	4	4 reset to x_1
48	22	**GTO**	
49	01	1	
50	03	3	To repeat with reduced Δx
51	.		Steps 51 through 99, plus registers 0, 6, 7, 8, and 9 available for constructing $f(x)$ from x, which is in display register at 51. End with **2nd rtn**
.			
.			
.			

23

Table 2-1. (HP-25) Definite integral by full-increment method

Register Contents:

0	1	2	3	4	5	6	7
—	x_1	x_2	Δx	x_i	Σ	—	—

(1 and 2 are preloaded)

Program

Step	Code		Key Entry	X	Y	Z	T	Comments
00								
01	24	02	RCL 2	x_2				
02	24	01	RCL 1	x_1	x_2			
03	23	04	STO 4	x_1	x_2			4 set to x_1
04		41	−	Δx				$\Delta x = x_2 - x_1$, initially
05	23	03	STO 3	Δx				
06		00	0	0				
07	23	05	STO 5	0				5 zeroed
08	24	02	RCL 2	x_2	x_2			
09	24	04	RCL 4	x_i	x_2			$x_i = x_1$, initially
10	14	51	f x ≥ y	x_i	x_2			Tests for x_i at end of interval
11	13	43	GTO 43	x_i	x_2			To display completed Σ

24

Step	Code	Keystroke	f(x_i)	Display	Comments
12					Steps 12 through 36, plus registers 0, 6, and 7 available for constructing f(x_i). If fewer steps used, end with **GTO 37**
.					
.					
.					
.					
.					
.					
36					
37	24 03	**RCL 3**	$f(x_i)$	$f(x_i)$	
38	61	**x**		Δx	
39	51 05	**STO + 5**		$f(x_i)\,\Delta x$	$f(x_i)\,\Delta x$ added to Σ
40	24 03	**RCL 3**		$f(x_i)\,\Delta x$	
41	51 04	**STO + 4**		Δx	x_i incremented
42	13 08	**GTO 08**		Δx	To evaluate next $f(x_i)\,\Delta x$
43	24 05	**RCL 5**		Σ	
44	74	**R/S**		Σ	
45	02	**2**		2	
46	71 03	**STO ÷ 3**		2	Δx reduced
47	24 01	**RCL 1**		x_1	4 reset to x_1
48	23 04	**STO 4**		x_1	
49	13 06	**GTO 06**		x_1	To repeat with reduced Δx

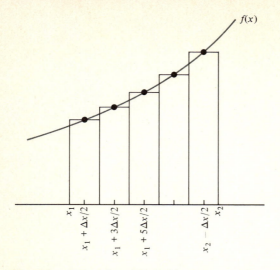

Figure 2-2. Illustration of the half-increment definition of a definite integral.

numerical-integration procedure. For a given size of Δx, the area under the rectangles whose tops straddle the curve is by far the better approximation to the area under the curve. The mathematical expression of the altered definition is

$$\int_{x_1}^{x_2} f(x)\,dx = \text{limit as } \Delta x \text{ approaches } 0 \text{ of } \left[\sum_{x_i} f(x_i)\,\Delta x\right]$$

where $x_i = x_1 + \Delta x/2,\ x_1 + 3\Delta x/2,\ \ldots,\ x_2 - \Delta x/2$.

Half-increment programs are listed in Tables 2-2 (SR-56) and (HP-25). They are used in exactly the same way as are the full-increment programs. For the same example considered before, they yield dramatically faster convergence.

Example

$f(x) = x^2,\ x_1 = 1,\ x_2 = 2$

[entered and run in the SR-56 or HP-25 exactly as for the full-increment programs, except for the difference in the steps available for constructing $f(x)$]. The results are:

$$\sum_{x_i} f(x_i)\,\Delta x = 2.25,\ 2.31,\ 2.33,\ 2.33$$

They converge to the same limit as do the results of the full-increment method applied to this example, and so are also in agreement with the predictions of the analytical method of integrating. But the half-increment method is rapid enough to be of real practical utility, as you will see in the next section.   ////

2-4. INTEGRATION OF THE SOLAR-RADIATION SPECTRUM

The rate of emission of radiation per unit wavelength interval by 1 m² (square meter) of the surface of the sun is (Ref. 2):

$$R(\lambda) = \frac{3.74 \times 10^{-16}}{\lambda^5 \left(e^{2.52 \times 10^{-6}/\lambda} - 1\right)} \qquad \text{W/m (watts per meter)}$$

where the wavelength λ is in meters. The integral

$$I = \int_{3.50 \times 10^{-7}}^{7.00 \times 10^{-7}} R(\lambda)\, d\lambda$$

measures the power radiated in the range of visible wavelengths. Despite its apparent simplicity the form of $R(\lambda)$ is such that its integral can be evaluated only by numerical methods.

But this is easy to do. Key into the empty region of the half-increment program appropriate to your calculator the entries listed in Tables 2-3 (SR-56) and (HP-25). These will construct $f(\lambda)$ with the aid of the constants 2.52×10^{-6} and 3.74×10^{-16} that you enter into the designated storage registers.

For the SR-56

You do this: With the calculator in the execute mode, key **GTO 5 8.** Then put it in the learn mode by keying **LRN.** Enter the subprogram by keying steps 58 through 80. Key **LRN** to return to execute mode, and **RST** to set to step 00. Load the constants by keying **3 . 7 4 EE 1 6 +/− STO 6**, and **2 . 5 2 EE 6 +/− STO 7.** Then you key **3 . 5 EE 7 +/− STO 1**, and **7 EE 7 +/− STO 2**, to load the limits of integration. Push **R/S** to start.

For the HP-25

You do this: With the calculator switched to **RUN,** key **GTO 1 4.** Then switch to **PRGM.** Enter the subprogram by keying in steps 15 through 29. Switch back to **RUN,** and set to step 00 by keying **f PRGM.** Load the constants by keying **3 . 7 4 EEX 1 6 CHS STO 6**, and **2 . 5 2 EEX 6 CHS STO 7.** Then you key **3 . 5 EEX 7 CHS STO 1**, and **7 EEX 7 CHS STO 2**, to load the limits of integration. It is convenient to then key **f SCI 2.** Push **R/S** to start.

When rounded off to two decimal places, the results are:

$I = 2.72 \times 10^7, 2.55 \times 10^7, 2.51 \times 10^7, 2.50 \times 10^7, 2.49 \times 10^7,$
2.49×10^7 **W**

Table 2-2. (SR-56) Definite integral by half-increment method

Register Contents:

0	1	2	3	4	5	6	7	8	9
—	x_1	x_2	Δx	x_i	Σ	—	—	—	—

preload (under registers 0, 1, 2)

Program

Step	Code	Key Entry	Comments	Step	Code	Key Entry	Comments
00	34	RCL		33	35	SUM	
01	02	2	x_2	34	05	5	$f(x_i)\,\Delta x$ added to Σ
02	32	$x \gtreqless t$	Test register loaded with x_2	35	34	RCL	
03	34	RCL		36	03	3	Δx
04	02	2	x_2	37	35	SUM	
05	74	−		38	04	4	x_i incremented
06	34	RCL		39	22	GTO	
07	01	1	x_1	40	02	2	
08	33	STO		41	01	1	To evaluate next $f(x_i)\,\Delta x$
09	04	4	4 set to x_1	42	34	RCL	
10	94	=	$\Delta x = x_2 - x_1$, initially	43	05	5	Σ
11	33	STO		44	41	R/S	
12	03	3		45	02	2	2
13	54	÷		46	12	INV	
14	02	2		47	30	2nd PROD	
15	94	=	$\Delta x/2$	48	03	3	Δx reduced

Step	Code	Key	Comment
16	35	**SUM**	
17	04	**4**	4 now initial x_i
18	00	**0**	0
19	33	**STO**	
20	05	**5**	5 zeroed
21	34	**RCL**	
22	04	**4**	$x_i = x_1 + \Delta x/2$, initially
23	47	**2nd** $x \geq$ **t**	Tests for x_i beyond interval
24	04	**4**	
25	02	**2**	
26	57	**2nd subr**	To evaluate $f(x_i)$
27	05	**5**	
28	08	**8**	
29	64	**×**	
30	34	**RCL**	
31	03	**3**	Δx
32	94	**=**	$f(x_i)\,\Delta x$

Step	Code	Key	Comment
49	34	**RCL**	
50	01	**1**	x_1
51	33	**STO**	
52	04	**4**	4 reset to x_1
53	34	**RCL**	
54	03	**3**	Δx
55	22	**GTO**	
56	01	**1**	
57	03	**3**	
58	. . .		To repeat with reduced Δx. Steps 58 through 99, plus registers 0, 6, 7, 8, and 9 available for constructing $f(x)$ from x, which is in display register at 58. End with **2nd rtn**

Table 2-2. (HP-25) Definite integral by half-increment method

Register Contents:

0	1	2	3	4	5	6	7
—	x_1	x_2	Δx	x_i	Σ	—	—
	preloaded						

Program

Step	Code	Key Entry	X	Y	Z	T	Comments
00							
01	24 02	RCL 2	x_2				
02	24 01	RCL 1	x_1	x_2			
03	23 04	STO 4	x_1	x_2			4 set to x_1
04	41	−	Δx				$\Delta x = x_2 - x_1$, initially
05	23 03	STO 3	Δx				
06	02	2	2	Δx			
07	71	÷	$\Delta x/2$				
08	23 51 04	STO + 4	$\Delta x/2$				4 now initial x_i
09	00	0	0				
10	23 05	STO 5	0				5 zeroed
11	24 02	RCL 2	x_2	x_2			
12	24 04	RCL 4	x_i	x_2			$x_i = x_1 + \Delta x/2$, initially
13	14 51	f x ≥ y	x_i	x_2			Tests for x_i beyond interval
14	13 42	GTO 42	x_i	x_2			To display completed Σ

Step	Code	Keystroke	Display	Comment
15				Steps 15 through 35, plus registers 0, 6, and 7 available for constructing $f(x_i)$. If fewer steps used, end with **GTO 36**
.				
.				
.				
.				
35				
36	24 03	RCL 3	$\dfrac{f(x_i)}{\Delta x}$ $f(x_i)$	
37	61	×	$f(x_i)\,\Delta x$	
38	23 51 05	STO + 5	$f(x_i)\,\Delta x$	$f(x_i)\,\Delta x$ added to Σ
39	24 03	RCL 3	Δx	
40	23 51 04	STO + 4	Δx	x_i incremented
41	13 11	GTO 11	Σ	To evaluate next $f(x_i)\Delta x$
42	24 05	RCL 5	Σ	
43	74	R/S	2	
44	02	2	2	
45	23 71 03	STO ÷ 3	x_1	Δx reduced
46	24 01	RCL 1	x_1	
47	23 04	STO 4	Δx	4 reset to x_1
48	24 03	RCL 3	Δx	
49	13 06	GTO 06		To repeat with reduced Δx

31

Table 2-3. (SR-56) Integration of solar spectrum

Register Contents:

0	1	2	3	4	5	6	7	8	9
λ_i	3.5×10^{-7}	7×10^{-7}	Δx	x_i	Σ	3.74×10^{-16}	2.52×10^{-6}	—	—
	preloaded					preloaded			

Program

Step	Code	Key Entry	Comments
58	33	STO	
59	00	0	0 loaded with λ_i
60	20	2nd 1/x	$1/\lambda_i$
61	64	×	
62	34	RCL	
63	07	7	$b = 2.52 \times 10^{-6}$
64	94	=	b/λ_i
65	14	e^x	e^{b/λ_i}
66	74	−	
67	01	1	1
68	94	=	$e^{b/\lambda_i} - 1$
69	64	×	

Step	Code	Key Entry	Comments
70	34	RCL	
71	00	0	λ_i
72	45	y^x	
73	05	5	
74	94	=	$\lambda_i^5(e^{b/\lambda_i} - 1)$
75	20	2nd 1/x	$1/\lambda_i^5(e^{b/\lambda_i} - 1)$
76	64	×	
77	34	RCL	
78	06	6	$a = 3.74 \times 10^{-16}$
79	94	=	$f(\lambda_i)$
80	58	2nd rtn	

Table 2-3. (HP-25) Integration of solar spectrum

Register Contents:

0	1	2	3	4	5	6	7
λ_i	3.5×10^{-7}	7×10^{-7}	Δx	x_i	Σ	3.74×10^{-16}	2.52×10^{-6}
	preloaded					preloaded	

Program

Step	Code	Key Entry	X	Y	Z	T	Comments
14							
15	23 00	**STO 0**	λ_i				0 loaded with λ_i
16	15 22	**g 1/x**	$1/\lambda_i$				
17	24 07	**RCL 7**	b	$1/\lambda_i$			$b = 2.52 \times 10^{-6}$
18	61	**×**	b/λ_i				
19	15 07	**g e^x**	e^{b/λ_i}				
20	01	**1**	1	e^{b/λ_i}			
21	41	**−**	$e^{b/\lambda_i} - 1$				
22	24 00	**RCL 0**	λ_i	$e^{b/\lambda_i} - 1$	$e^{b/\lambda_i} - 1$		
23	05	**5**	5	λ_i	$e^{b/\lambda_i} - 1$		
24	14 03	**f y^x**	λ_i^5	$e^{b/\lambda_i} - 1$			
25	61	**×**	$(e^{b/\lambda_i} - 1)\lambda_i^5$				
26	15 22	**g 1/x**	$1/((e^{b/\lambda_i} - 1)\lambda_i^5)$				
27	24 06	**RCL 6**	a	$1/((e^{b/\lambda_i} - 1)\lambda_i^5)$			$a = 3.74 \times 10^{-16}$
28	61	**×**	$f(\lambda_i)$				
29	13 36	**GTO 36**	$f(\lambda_i)$				

EXERCISES

2-1. Run the full- and half-increment programs to evaluate the integral of $f(x) = x^2$ at several different sets of values of x_1 and x_2. For each set, compare the speeds of convergence of the two programs.

2-2. Try other functions such as power laws, sinusoidals, and exponentials in the two programs. For what functions is the discrepancy in speed of convergence between the two programs most striking? Why?

2-3. Use the half-increment program to test your knowledge of the analytical expression for the definite integral over the range from some x_1 to some x_2 of a variety of functions. Do this by running them in the program and then comparing the results with what you get by using the calculator to evaluate the analytical expression.

2-4. Run the half-increment program to evaluate the definite integral over some range for a function for which an analytical expression does not exist. A particularly important example is the normal probability integral

$$\frac{1}{\sqrt{2\pi}} \int_{-x_1}^{x_1} e^{-x^2/2} dx$$

which plays an important role in the error analysis of experimental data. Using several different values of x_1, make your own table of the values of this integral and compare these with the values that can be found in any table of mathematical data.

2-5. Using the half-increment program to evaluate the definite integral of some function from $x_1 = 0$ to a number of uniformly spaced values of x_2, make a plot of its definite integral versus x_2. Then, using the calculator to evaluate the function, plot it and compare it with the plot of its definite integral. Interesting functions to consider are $\sin x$ and $\cos x$.

2-6. Explain why in the integration programs, in contrast to the differentiation programs, you never have to worry about calculator roundoff error.

2-7. Study the details of each program and explain precisely what happens in each step.

2-8. There is another numerical integration procedure that converges rapidly enough to be useful and is simple enough to run on a programmable pocket calculator. It amounts to calculating the total area under a set of trapezoids whose sides are perpendicular to the x axis and intersect it at $x_1, x_1 + \Delta x, \ldots, x_2$, and whose tops are straight lines joining the intersections of the sides and the curve $f(x)$. Show that the total area will be:

$$\frac{f(x_1)}{2}\Delta x + f(x_1 + \Delta x)\Delta x + f(x_1 + 2\Delta x)\Delta x + \cdots$$
$$+ f(x_2 - \Delta x)\Delta x + \frac{f(x_2)}{2}\Delta x$$

Then write a program to carry out this trapezoidal procedure and compare its results with those of the half-increment program for $f(x) = x^2$, $x_1 = 1$, $x_2 = 2$.

REFERENCES

1. Bueche, Frederick J.: *Introduction to Physics for Scientists and Engineers,* 2d ed., McGraw-Hill Book Company, New York, 1975, pp. 841–844.
   Halliday, David, and Robert Resnick: *Fundamentals of Physics,* John Wiley & Sons, Inc., New York, 1974, p. 100.
   Stein, Sherman K.: *Calculus and Analytic Geometry,* McGraw-Hill Book Company, New York, 1973, p. 248.
2. Eisberg, Robert, and Robert Resnick: *Quantum Physics of Atoms, Molecules, Solids, Nuclei, and Particles,* John Wiley & Sons, Inc., New York, 1974, p. 21.
   Halliday, David, and Robert Resnick: *Fundamentals of Physics,* John Wiley & Sons, Inc., New York, 1974, p. 761.

THREE

NUMERICAL SOLUTION OF DIFFERENTIAL EQUATIONS

3-1. INTRODUCTION

You now come to the heart of this book—solving differential equations important to elementary physics on a programmable pocket calculator. It is a lot easier than it sounds. There will be no assumption that you know anything about analytical methods for solving differential equations, or even that you know what a differential equation is. The only math prerequisite is that you know what a derivative is. Specifically, if a particle moving along a straight line is at position y at time t, then you should know that dy/dt is the time rate of change of y, or its *speed*, and that $d^2y/dt^2 \equiv d(dy/dt)/dt$ is the time rate of change of its speed dy/dt, or its *acceleration*. If you are not familiar with derivatives, look at Chap. 1.

The only physics prerequisite for this chapter and the three that follow is that you know about *Newton's law of motion:*

$$F = m \frac{d^2y}{dt^2}$$

which relates the *force F* acting on a particle to its *mass m* and its acceleration d^2y/dt^2. If this is unfamiliar, consult one of the books in Ref. 1 at the end of this chapter.

You will start by considering a very simple physical system: A particle falling without friction, under the constant gravitational force it feels near the surface of the earth. As applied to this case, the numerical method is as rudimentary as it can be, and even the analytical method is so trivial that its predictions can be obtained by a very short argument. This will allow you to compare the results of the numerical and analytical methods, thereby gaining confidence in the numerical method and your ability to handle it.

Actually, two numerical methods will be developed. You guessed it! There is a full-increment method and a half-increment method. The latter is only a little more complicated than the former, but very much more accurate. Furthermore, it is of very wide applicability. At the end of this chapter it will be applied to a more interesting problem involving fall near the surface of the earth: An object falling under a constant gravitational force, but experiencing also a frictional force of strength proportional to the square of its speed. A concrete example that will be treated is the case of a skydiver.

In the three chapters that follow, the half-increment method will be applied to a wide variety of mechanical systems that are important in classical physics. You will see that the method employed to solve any one of the differential equations that describe the systems is very much the same as the method employed to solve the others, despite the fact that the equations are sufficiently different that their analytical treatments vary strikingly from one to the other. Furthermore, for some of the important systems that you will learn to treat numerically, there is *no* analytical solution to the differential equations. Herein lies the great advantage of the numerical method. It almost always works, and in almost exactly the same way!

Many of the differential equations you will solve in these four chapters arise in exactly the same mathematical form in fields of classical physics other than mechanics. If you are interested in electromagnetism, you certainly will be able to find many useful applications of the physics to be learned from studying the behavior of the solutions.

In the final chapter the half-increment method will be applied, with an interesting new twist, to solve Schroedinger's equation. This equation plays the role in quantum mechanics that Newton's law plays in classical mechanics. The chapter is written so as to make the least possible demands on your prior background in physics.

3-2. FULL-INCREMENT SOLUTION FOR FREE FALL

All the objects studied in classical mechanics obey Newton's law of motion, $F = m \, d^2y/dt^2$, which is a differential equation because it contains a derivative. It is most convenient to write it as

$$\frac{d^2y}{dt^2} = \frac{F}{m} \tag{3-1}$$

If you can find the form of the force F, and you know the mass m, then you can always determine the behavior of the object by using numerical methods to solve the equation and obtain its position y at closely spaced values of the time t.

When unsupported near the surface of the earth, and neglecting friction, the only force F a body feels is the gravitational attraction of the earth; i.e., the force which makes it have weight. Since experiment shows its weight is proportional to its mass, you can write

$$F = gm$$

where g is some proportionality constant. Substituting into Eq. (3-1) gives

$$\frac{d^2y}{dt^2} = \frac{gm}{m} = g \tag{3-2}$$

where, for simplicity, the downward direction is taken to be the direction of positive y. See Fig. 3-1. The mass independence of the acceleration d^2y/dt^2 certainly agrees with experiment. In fact, measurements you may have seen made, or made yourself, provide the following value of the proportionality constant, which is the *gravitational acceleration*

$$g = 9.8 \text{ m/s}^2 \text{ (meters per second squared)} \tag{3-3}$$

Now to develop a numerical method for solving the differential equation (3-2). First rewrite it as

$$\frac{d}{dt} \frac{dy}{dt} = g$$

Then evoke the full-increment definition of a derivative in Sec. 1-2 as applied to a case in which the quantity being differentiated is dy/dt. It yields the approximate result

$$\frac{d}{dt} \frac{dy}{dt_i} \simeq \frac{\dfrac{dy}{dt_{i+1}} - \dfrac{dy}{dt_i}}{\Delta t}$$

where the equality is the more accurate the smaller the value of Δt, where the subscript i means the entire term preceding it is evaled at some initial time, and where the subscript $i + 1$ means it is evaled at a time which is later by the amount Δt. Using the previous equation you have

$$\frac{\dfrac{dy}{dt_{i+1}} - \dfrac{dy}{dt_i}}{\Delta t} \simeq g$$

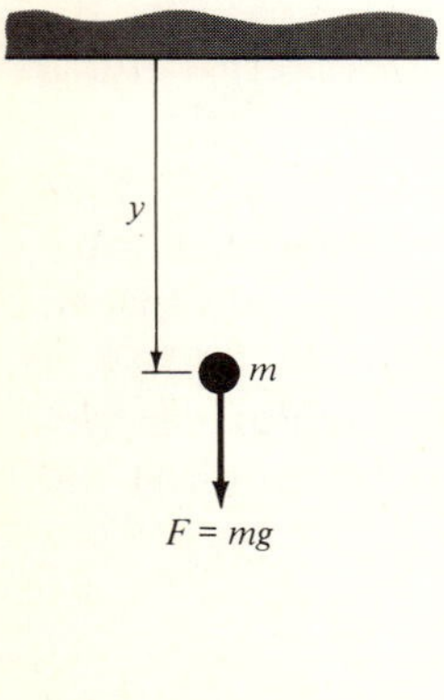

Figure 3-1. A body of mass m falling vertically downward under the influence of gravitational force $F = mg$. If there is also a frictional force, it acts in the direction that will oppose the motion of the body, or upward in this case.

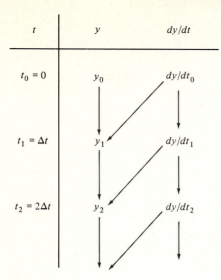

t	y	dy/dt
$t_0 = 0$	y_0	dy/dt_0
$t_1 = \Delta t$	y_1	dy/dt_1
$t_2 = 2\Delta t$	y_2	dy/dt_2

Figure 3-2. Scheme for the full-increment solution of the differential equation for free fall.

No subscript is needed on the g, of course, since it is a constant. The next step gives one of the equations basic to the method:

$$\frac{dy}{dt_{i+1}} \simeq \frac{dy}{dt_i} + g\,\Delta t \qquad (3\text{-}4)$$

The other equation comes directly from the full-increment definition of a derivative:

$$\frac{dy}{dt_i} \simeq \frac{y_{i+1} - y_i}{\Delta t}$$

where the subscripts i and $i+1$ have the previous meanings. This gives

$$y_{i+1} \simeq y_i + \frac{dy}{dt_i}\Delta t \qquad (3\text{-}5)$$

Equations (3-4) and (3-5) form the basis of a numerical-solution method which can be most easily understood by referring to Fig. 3-2. The scheme indicated in the figure shows how Eqs. (3-4) and (3-5) can be used to determine approximately the position y and speed dy/dt of the object at any of a closely spaced set of subsequent values of the time t, *if* its position y_0 and speed dy/dt_0 are known at an initial value of time t_0 which, for simplicity, is taken to have the value $t_0 = 0$. It goes as follows. First, y_0 and dy/dt_0 are used in (3-5) to determine y_1. Then dy/dt_0 and the value of g are used in (3-4) to determine dy/dt_1. This completes the first loop of the method, which has produced approximate values for y_1 and dy/dt_1. The second loop is carried out in exactly the same manner and produces approximate values of y_2 and dy/dt_2 from the values of y_1, dy/dt_1, and g.

Table 3-1. (HP-25) Full-increment solution for free fall

Register Contents:

0	1	2	3	4	5	6	7
—	Δt	g	y	dy/dt	t	—	—
	preloaded						

Program

Step	Code	Key Entry	X	Y	Z	T	Comments
00	00	0	0				
01	23 03	STO 3	0				3 zeroed
02	23 04	STO 4	0				4 zeroed
03	23 05	STO 5	0				5 zeroed
04	24 01	RCL 1	Δt				Start loop
05	23 51 05	STO + 5	Δt				5 now t_1
06	24 04	RCL 4	dy/dt	Δt			
07	61	x	$\Delta t\, dy/dt$				
08	23 51 03	STO + 3	$\Delta t\, dy/dt$				3 now y_1
09	24 03	RCL 3	y				
10	14 74	f PAUSE	y				
11	(or 74)	(or R/S)					
12	24 02	RCL 2	g				
13	24 01	RCL 1	Δt	g			
14	61	x	$g\,\Delta t$				
15	23 51 04	STO + 4	$g\,\Delta t$				4 now dy/dt_1
16	13 05	GTO 05	$g\,\Delta t$				To new loop

The programs in Tables 3-1 (HP-25) and (SR-56) implement this method. They assume fall from rest, i.e., that $dy/dt_0 = 0$, and also that y is measured from the point where fall begins, i.e., that $y_0 = 0$. To run them, proceed with the following instructions.

For the HP-25
Switch to **PRGM,** clear by keying **f PRGM,** then key in steps 01 through 10. If you want to make a graph of the results for y at uniformly spaced values of t, key **R/S** in step 11. (When graphing, plot the points directly from the displayed numbers.) If you only want to watch y, key **f PAUSE** in that step. Then key the remainder of the steps in the program. Switch to **RUN,** and set the calculator to step 00 by keying **f PRGM.** Then load Δt by keying the value you choose, followed by **STO 1,** and, if you are using the metric system (SI), load g by keying **9 . 8 STO 2.** To start, push **R/S.** If step 11 is **R/S,** the calculator will stop after obtaining each new value of y and wait for you to restart it by pushing **R/S** again. Otherwise, it will only pause for about one second to display y, but can be stopped during a pause if you push **R/S.** If you want to see the value of t corresponding to a displayed y, key **RCL 5** while the calculator is stopped and displaying y. This will not confuse subsequent calculations.

For the SR-56
Clear by keying **2nd CP,** key **LRN,** then key in steps 00 through 18. If you want to make a graph of the results for y at uniformly spaced values of t, key **R/S** in step 19 and **2nd NOP** in step 20. (When graphing, plot the points directly from the displayed numbers.) If you only want to watch y, key **2nd pause** in both steps. Then key the remainder of the steps in the program. Return to execute mode by keying **LRN,** and set the calculator to step 00 by keying **RST.** Then load Δt by keying the value you choose, followed by **STO 1,** and, if you are using the metric system (SI), load g by keying **9 . 8 STO 2.** To start, push **R/S.** If step 19 is **R/S,** the calculator will stop after obtaining each new value of y and wait for you to restart it by pushing **R/S** again. In this mode you can see t when stopped by keying **RCL 5** before you restart. If step 19 is a "pause" instruction, the calculator will do so for about one second to display, but can be stopped after a pause by pushing **R/S.** If you then want to see t, key **2nd EXC 5,** inspect it, then key **2nd EXC 5** again. Neither procedure for seeing t will confuse subsequent calculations.

Before considering an example, it is appropriate to go through an analytical argument leading to a prediction of the dependence of y on t so that you will have something to compare to the results of the numerical method. As advertized, the argument is so trivial that it can be presented effectively without the aid of equations. Since the acceleration of an ob-

Table 3-1. (SR-56) Full-increment solution for free fall

Register Contents:

0	1	2	3	4	5	6	7	8	9
—	Δt	g	y	dy/dt	t	—	—	—	—
	preloaded								

Program

Step	Code	Key Entry	Comments
00	00	0	0
01	33	STO	
02	03	3	3 zeroed
03	33	STO	
04	04	4	4 zeroed
05	33	STO	
06	05	5	5 zeroed
07	34	RCL	Start loop
08	01	1	Δt
09	35	SUM	
10	05	5	5 now t_1
11	64	×	
12	34	RCL	
13	04	4	dy/dt
14	94	=	$\Delta t\, dy/dt$
15	35	SUM	
16	03	3	3 now y_1

Step	Code	Key Entry	Comments
17	34	RCL	
18	03	3	y_1
19	59 (or 41)	2nd PAUSE (or R/S)	
20	59 (or 46)	2nd PAUSE (or 2nd NOP)	
21	34	RCL	
22	02	2	g
23	64	×	
24	34	RCL	
25	01	1	Δt
26	94	=	$g\Delta t$
27	35	SUM	
28	04	4	4 now dy/dt_1
29	22	GTO	
30	00	0	
31	07	7	To new loop

ject falling freely from rest has the constant value g, when the elapsed time is t its speed has the value gt. And since its speed is increasing uniformly from zero, the average speed over the time interval ending at t is $gt/2$. As it is always true that the total distance covered is the average speed multiplied by the elapsed time, it follows that when the elapsed time is t, the value of y will be $(gt/2)t$, or

$$y = \frac{gt^2}{2} \tag{3-6}$$

Example:

$\Delta t = 0.5$, $g = 9.8$

Results obtained from running either calculator program with these values for Δt and g are shown by points in Fig. 3-3. Since the value of g used is in units of meters per second squared, the numbers on the t axis are in seconds and those on the y axis are in meters. The crosses are obtained by evaluating the analytical prediction of Eq. (3-6) at a few values of t. Comparison shows that the results obtained by the numerical solution of the differential equation are not too satisfactory. However, they would improve if a smaller value of Δt were used, just as the accuracy of the full-increment approximation for a derivative that the method is based on will improve as Δt is reduced.   ////

It is not feasible to do too much in the way of reducing Δt, and certainly not feasible to do a set of runs for successively smaller values of Δt and watch the results of the full-increment method converge to the analytical results. They *will* so converge. But for each Δt a complete run must be made, and it is necessarily the case that every reduction in Δt by a factor of $\frac{1}{2}$ increases the time required by a factor of 2, since it doubles the number of increments used to reach a given value of t. It is better, instead, to try to understand the reason for the method's inaccuracy and then to use that understanding to suggest a better method.

The inaccuracy arises from the fact that the method is based on the full-increment definition of a derivative. Look at Fig. 1-1. Note that the way the definition is used here amounts to taking the value of dy/dt at the beginning of a time increment Δt and employing it to estimate the change in y that occurs in that increment. For a y-versus-t curve that is concave upwards, as in both that figure and in Fig. 3-3, the result will be to underestimate the increase in y, since the slope of the curve dy/dt, which is the rate of increase of y, is always smallest at the beginning of the increment. Furthermore, the error arising from the underestimate is cumulative from each increment to the next. These considerations suggest a very simple solution—use the half-increment definition of a derivative.

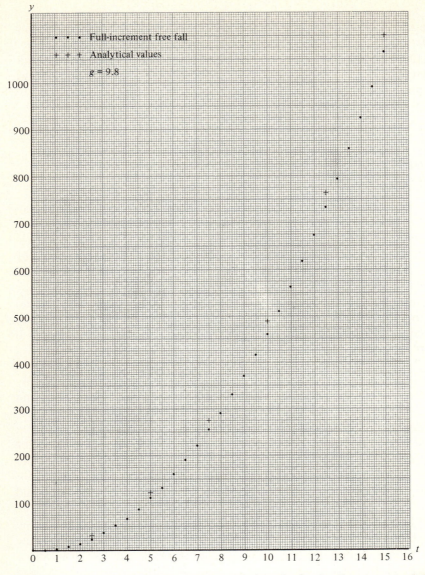

Figure 3-3. Analytical and full-increment numerical results for the distance *y* traveled in time *t* by a body falling freely from rest. With the gravitational acceleration *g* in meters per second squared, *y* is in meters and *t* in seconds.

3-3. HALF-INCREMENT SOLUTION FOR FREE FALL

It is not necessary to repeat, with modifications dictated by the difference between the half- and full-increment definitions of a derivative, the derivation leading to Eqs. (3-4) and (3-5). If you understand the basis and disadvantages of the full-increment method, you will immediately understand the explanation in the next paragraph of the basis and advantages of the following equations, and the scheme for using them shown in Fig. 3-4:

$$\frac{dy}{dt}_{1/2} \simeq \frac{dy}{dt}_0 + g\frac{\Delta t}{2} \tag{3-7}$$

$$\frac{dy}{dt}_{i+1/2} \simeq \frac{dy}{dt}_{i-1/2} + g\,\Delta t \tag{3-8}$$

$$y_{i+1} \simeq y_i + \frac{dy}{dt}_{i+1/2}\,\Delta t \tag{3-9}$$

The method involves first using dy/dt_0 and g to calculate the approximate value of $dy/dt_{1/2}$, the derivative at half a time increment later than the initial time. This is done by using Eq. (3-7), which is just (3-4) with Δt changed to $\Delta t/2$ and the subscripts then suitably modified. Then Eq. (3-9), which is identical to (3-5) except for subscripts, is used to produce an approximation to y_1 from y_0 and $dy/dt_{1/2}$. This completes the preliminary loop of the calculation. The next loop, and all subsequent ones, uses Eq. (3-8), which is identical to (3-4) except for subscripts, and Eq. (3-9) in the scheme in exactly the same way that (3-4) and (3-5) are used in the full-increment scheme of Fig. 3-2. The half-increment method is very much

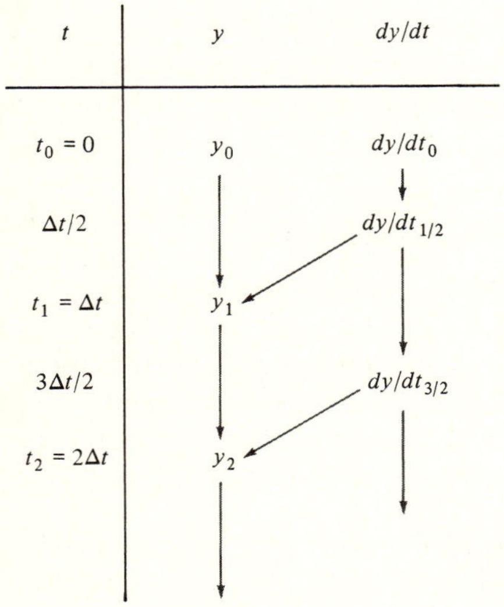

t	y	dy/dt
$t_0 = 0$	y_0	dy/dt_0
$\Delta t/2$		$dy/dt_{1/2}$
$t_1 = \Delta t$	y_1	
$3\Delta t/2$		$dy/dt_{3/2}$
$t_2 = 2\Delta t$	y_2	

Figure 3-4. Scheme for the half-increment solution of the differential equation for free fall.

more accurate than the full-increment one since the change in y in an increment of t is approximated by using the value of dy/dt at the midpoint of that increment.

Programs for the half-increment method are listed in Tables 3-2 (HP-25) and (SR-56). They are entered and run in precisely the same way as the corresponding full-increment programs and they also take $y_0 = dy/dt_0 = t_0 = 0$.

Example

$\Delta t = 0.5$, $g = 9.8$

The results obtained with these values are shown as points in Fig. 3-5, along with crosses representing the analytical results for a few values of t. The agreement looks very good indeed. If you run the half-increment program for your calculator and compare the actual numbers produced with those obtained by using the calculator to evaluate the analytical prediction of Eq. (3-6), you will see that the two methods agree perfectly. ////

Before you become too enthusiastic about the half-increment method, it should be pointed out that it is not typical for it to be *perfectly* accurate. The reason why the half-increment method is error free, for any value of Δt you happen to use, in the special situation of free fall has to do with a relation particular to that situation that was used in the argument leading to Eq. (3-6). As a self-test, see if you can explain the reason.

But it certainly is true that in all situations the half-increment method of solving differential equations is far superior to the full-increment method because for a given size of the increment in t (or in whatever independent variable enters in the equation) the results are much more accurate. To put the matter another way, for a given amount of running time, the half-increment method produces much more accurate results. As a consequence, it is the only method that henceforth will be used in this book.

3-4. EXPLANATION OF PROGRAM OPERATION

The program in either version of Table 3-2 is the most important one in the book because it can act as a prototype for all differential-equation solving programs. Although there are some differences in detail, all the differential equations you will encounter in this book and many, many more can be solved by equations, a scheme, and a program which are in spirit just like Eqs. (3-7), (3-8), (3-9), Fig. 3-4, and Tables 3-2 (HP-25) or (SR-56).

Therefore, it is appropriate to give you a detailed explanation of the operation of both versions of the program, just as is done in Sec. 1-3 for a program that is important because it is the first to be presented. The explanation will assume that you have read that section.

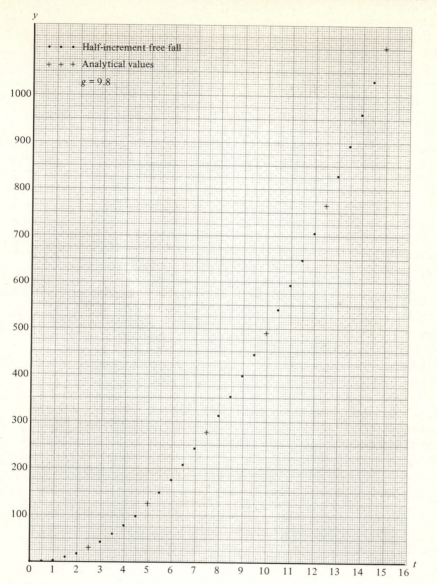

Figure 3-5. Analytical and half-increment numerical results for a body falling freely from rest.

Table 3-2. (HP-25) Half-increment solution for free fall

				Register Contents:				
0	1	2	3	4	5	6	7
—	Δt	g	y	dy/dt	t	—	—
	preloaded						

Program

Step	Code	Key Entry	X	Y	Z	T	Comments
00							
01	00	0	0				
02	23 03	STO 3	0				3 zeroed
03	23 04	STO 4	0				4 zeroed
04	23 05	STO 5	0				5 zeroed
05	24 02	RCL 2	g				Start loop
06	24 05	RCL 5	t	g			
07	15 71	g $x = 0$	t	g			Test t for routing to put 1/2 in preliminary loop
08	13 21	GTO 21	t	g			
09	22	R↓	g	g			Will be $g/2$ in prelim. loop
10	24 01	RCL 1	Δt	g			
11	61	x	$g\Delta t$				4 now $dy/dt_{1/2}$ in prelim. loop
12	23 51 04	STO + 4	$g\Delta t$				4 now $dy/dt_{3/2}$ in next loop

48

Step		Code	Keystrokes			Comment
13		24 01	RCL 1	Δt	Δt	5 now t_1 in prelim. loop
14	23	51 05	STO + 5	Δt		5 now t_2 in next loop
15		24 04	RCL 4	dy/dt		
16	23	61	x	Δt dy/dt		
17		51 03	STO + 3	Δt dy/dt		
18		24 03	RCL 3	y		3 now y_1 in prelim. loop
19		14 74 (or 74)	f PAUSE (or R/S)	y		3 now y_2 in next loop
20		13 05	GTO 05	y		To new loop
21		22	R↓	g	g	
22		02	2	2		
23		71	÷	g/2		
24	13	10	GTO 10	g/2		To continue loop

Table 3-2. (SR-56) Half-increment solution for free fall

Register Contents:

0	1	2	3	4	5	6	7	8	9
—	Δt	g	y	dy/dt	t	$g, g/2$	—	—	—
	preloaded								

Program

Step	Code	Key Entry	Comments	Step	Code	Key Entry	Comments
00	00	0	0	26	01	1	Δt
01	33	STO		27	35	SUM	
02	03	3	3 zeroed	28	05	5	5 now t_1 in prelim. loop, t_2 in next
03	33	STO		29	64	×	
04	04	4	4 zeroed	30	34	RCL	$dy/dt_{1/2}$ in prelim. loop, $dy/dt_{3/2}$ in next
05	33	STO		31	04	4	
06	05	5	5 zeroed	32	94	=	$\Delta t\, dy/dt_{1/2}$ in prelim. loop, $\Delta t\, dy/dt_{3/2}$ in next
07	56	2nd CP	Test register zeroed	33	35	SUM	
08	34	RCL	Start loop	34	03	3	3 now y_1 in prelim. loop, y_2 in next
09	02	2	g	35	34	RCL	
10	33	STO		36	03	3	
11	06	6	6 now g	37	59	2nd PAUSE	y_1 in prelim. loop, y_2 in next
12	34	RCL			(or 41)	(or R/S)	
13	05	5	t				
14	37	2nd x = t	Test t for routing in preliminary loop				
15	04	4					

Step	Code	Instruction	Comment
16	02	2	
17	34	RCL	
18	01	1	$g/2$ in prelim. loop, g in next loop
19	64	×	
20	34	RCL	
21	06	6	Δt
22	94	=	$\Delta t g/2$ in prelim. loop, $\Delta t g$ in next loop
23	35	SUM	
24	04	4	4 now $dy/dt_{1/2}$ in prelim. loop, $dy/dt_{3/2}$ in next loop
25	34	RCL	

Step	Code	Instruction	Comment
38	59 (or 46)	2nd PAUSE (or 2nd NOP)	
39	22	GTO	
40	00	0	
41	08	8	To new loop
42	02	2	2
43	12	INV	
44	30	2nd PROD	
45	06	6	6 now $g/2$
46	22	GTO	
47	01	1	
48	07	7	To continue loop

For the HP-25

Step 01 puts the digit 0 in the X register of the stack so that steps 02, 03, and 04 can use it to set registers 3, 4, and 5 to zero before the looping starts. Step 05 recalls the contents of 2, namely g, and puts it in X where it will be used later. Step 06 puts t in X by recalling it from 5, thereby displacing g up to the Y register of the stack. To identify the current value of t, the test in step 07 asks the question: Does $t = 0$?. If the answer is *yes*, as it will be in the preliminary loop run through first while $dy/dt_{1/2}$, t_1, and y_1 are being obtained, the calculator goes to the next step. In that step, 08, the calculator is routed to step 21. As it reaches 21, it still has t in X and g in Y. Step 21 makes the stack "roll down," thereby putting g in X. Steps 22 and 23 produce $g/2$ in X by first placing the digit 2 in X, which raises g to Y, and then dividing the contents of Y by the contents of X. The next step, 24, returns the calculator to step 10, carrying $g/2$ in X.

Step 10 recalls Δt from 1, pushing $g/2$ in the preliminary loop (or g in the next loop) to Y. In step 11 multiplication produces $(g/2)\,\Delta t$ in the preliminary loop (or $g\,\Delta t$ in the next loop) in X, which is added to the contents of register 4 in step 12. Thus 4 will now contain $dy/dt_{1/2}$ in the preliminary loop (or $dy/dt_{3/2}$ in the next loop). Steps 13 and 14 recall Δt from register 1 so that it can be added to 5, thereby making its contents t_1 in the preliminary loop (or t_2 in the next loop). Step 15 recalls register 4, thereby placing in X the quantity $dy/dt_{1/2}$ in the preliminary loop (or $dy/dt_{3/2}$ in the next loop), and also pushing the Δt that had been there up to Y. In steps 16 and 17, multiplication produces $\Delta t\,dy/dt_{1/2}$ in the preliminary loop (or $\Delta t\,dy/dt_{3/2}$ in the next loop), which is added to the contents of register 3. Thus that register now contains y_1 in the preliminary loop (or y_2 in the next loop). Steps 18 and 19 recall the new value of y to the always displayed X register and make the calculator pause, or stop, so that it can be inspected.

After the pause, or restart, step 20 makes the calculator return to step 05, where it begins the next loop. When it reaches the test in step 07, the answer to the question posed will this time (and in all subsequent loops) be *no*. So the calculator will skip step 08 and, thereby, not be routed to the steps at the end of the program that insert the factor of $\frac{1}{2}$ needed only in the preliminary loop. Thus after the roll down called for in step 09, the contents of X will be g instead of $g/2$. From here on, this loop and all subsequent loops continue as before.

For the SR-56

Step 00 puts the digit 0 in the display register, and steps 01 through 06 use it to "zero" registers 3, 4, and 5. Step 07 sets to zero a test register. Steps 08 and 09 recall g from 2 into the display register. Steps 10 and 11 store it in register 6. In steps 12 and 13, t is recalled from 5 into the display register. Step 14 compares t with the value zero earlier stored in the test register. Specifically, step 14 asks the question: Is the content of the display register equal to the content of the test register? If the answer is *yes* to this question, or to any of the other test questions available to the calculator, then it will go to the step specified by the digits in the two steps immediately below the test question. So, since the answer will be yes in the preliminary loop in which $dy/dt_{1/2}$, t_1, and y_1 are being obtained, the calculator will go to step 42.

That step puts the digit 2 in the display register so that steps 43, 44, and 45 can divide it into the contents of register 6. So register 6 now contains $g/2$. Steps 46, 47, and 48 send the calculator back to step 17.

This step, and step 18, recall Δt from 1 and put it in the display register. Step 19 prepares a multiplication. Steps 20 and 21 recall the contents of 6 into the display register. This will be $g/2$ in the preliminary loop (or g in the next loop). So when the multiplication is completed by step 22, the display register will contain $\Delta t\, g/2$ in the preliminary loop (or $\Delta t\, g$ in the next loop). Steps 23 and 24 add this quantity to the contents of register 4, which will now contain $dy/dt_{1/2}$ in the preliminary loop (or $dy/dt_{3/2}$ in the next loop). Steps 25 through 28 recall Δt from 1 into the display register and then add it into the contents of 5, which now contains t_1 in the preliminary loop (or t_2 in the next loop). But Δt remains in the display register so it can be prepared for a multiplication by step 29. Steps 30 and 31 recall the contents of register 4, which will be $dy/dt_{1/2}$ in the preliminary loop (or $dy/dt_{3/2}$ in the next loop). The multiplication is completed in step 32, producing $\Delta t\, dy/dt_{1/2}$ in the preliminary loop (or $\Delta t\, dy/dt_{3/2}$ in the next loop). Steps 33 and 34 add that number into register 3, which now contains y_1 in the preliminary loop (or y_2 in the next loop). Steps 35 through 38 recall it to the display register, make the calculator pause, or stop, and unblank the display.

After the pause, or restart, steps 39, 40, and 41 make the calculator return to step 08 where it begins the next loop. When it reaches step 14, the answer to the question posed will this time (and in all subsequent loops) be *no*. When the answer to this or any other test question is no, the calculator goes to the third step below the test question. So it will go to step 17 and, thereby, not be routed to the steps at the end of the program that insert the factor of $\frac{1}{2}$ needed only in the preliminary loop. Thus when 6 is recalled in step 21, the contents of the display register will be g instead of $g/2$. From here on, this loop and all subsequent loops continue as before.

3-5. FALL WITH FRICTION PROPORTIONAL TO $(dy/dt)^2$

Real objects falling through the air experience a frictional-retarding force due to their air resistance. For small objects moving slowly, experiment shows that the magnitude of the *frictional force* is approximately proportional to the first power of their speed, while it is approximately proportional to the square of their speed for large objects moving rapidly (Ref.

2). An example of the latter, which will be treated in this section by the half-increment method, is the fall of a skydiver.

It is easy to write the differential equation governing the motion of a skydiver falling with friction proportional to $(dy/dt)^2$. Starting from Newton's law of motion

$$\frac{d^2y}{dt^2} = \frac{F}{m}$$

and expressing the total force F as the gravitational force mg minus the frictional force $f(dy/dt)^2$, for the reason indicated in the legend to Fig. 3-1, the differential equation

$$\frac{d^2y}{dt^2} = \frac{mg - f\left(\frac{dy}{dt}\right)^2}{m}$$

emerges immediately. The proportionality constant f governs the overall strength of the frictional force and depends on the size and shape of the object and the viscosity of the material it is falling through. It is convenient to rewrite the equation as

$$\frac{d^2y}{dt^2} = g - \alpha\left(\frac{dy}{dt}\right)^2 \tag{3-10}$$

where

$$\alpha = \frac{f}{m} \tag{3-11}$$

It is almost as easy to modify the numerical method used previously so that it can handle this differential equation. The first thing is to subtract $\alpha\,(dy/dt)^2$ from g in Eqs. (3-7), (3-8), and (3-9), producing

$$\frac{dy}{dt}_{1/2} \simeq \frac{dy}{dt_0} + C\,\frac{\Delta t}{2} \qquad C = g - \alpha\left(\frac{dy}{dt}\right)_0^2 \tag{3-12}$$

$$\frac{dy}{dt}_{i+1/2} \simeq \frac{dy}{dt_{i-1/2}} + C\,\Delta t \qquad C = g - \alpha\left(\frac{dy}{dt}\right)_{i-1/2}^2 \tag{3-13}$$

$$y_{i+1} \simeq y_i + \frac{dy}{dt}_{i+1/2}\,\Delta t \tag{3-14}$$

The scheme for using these equations remains as before. The program must be modified but, as inspection of Tables 3-3 (HP-25) or (SR-56) will show, all that is involved is the addition of four or five very straightforward steps.

For the HP-25
The program is entered and run just as the preceding program, except that α must be loaded by keying its value followed by **STO 3**. If you want to see the value of dy/dt, or of t, corresponding to a displayed y, key **RCL 5**, or **RCL 6**, while the calculator is stopped to display y.

For the SR-56

The program is entered and run just as the preceding program, except that α must be loaded by keying its value followed by **STO 3.** You can see the value of dy/dt, or of t, corresponding to a displayed y. Just follow the **RCL 5,** or **6,** or else the double **2nd EXC 5,** or **6,** procedure described before.

It is not at all easy to change the analytical argument so that it will be able to take square-law friction properly into account. That is why this interesting physical system is not treated or even mentioned in most elementary physics textbooks. If you look at the results of the analytical method (Ref. 3)

$$y = \frac{1}{\alpha} \ln \left[\frac{e^{(\alpha g)^{1/2} t} + e^{-(\alpha g)^{1/2} t}}{2} \right] \tag{3-15}$$

you may get some idea of the complexity of the mathematical analysis required to derive it. This formula is quoted, without proof, because it will be useful in making a second test of the accuracy of the half-increment method of solving differential equations.

Example

$\Delta t = 0.5$, $g = 9.8$, $\alpha = 0.003$

The results obtained by running either version of the program with these parameters are plotted as points in Fig. 3-6, which shows both y and dy/dt. Using the value 9.8 for g means that the units of y are in meters, dy/dt in meters per second, and t in seconds. Also, it is easy to see from Eq. (3-10) that the units of α must be reciprocal meters so that $\alpha (dy/dt)^2$ will have the same units as g. The value of α was chosen to represent the actual situation for a skydiver.

Note how the skydiver's speed dy/dt builds up rapidly for the first few seconds of the jump, but soon plateaus at a value around 60 m/s. This terminal speed can be adjusted if desired by extending or retracting arms and legs to change α. Also note that the distance y the skydiver falls versus the time t is roughly parabolic (as in Fig. 3-5) while speed is relatively low, but becomes linear with the approach of terminal speed.

The crosses represent the results of the analytical method for y and can be obtained by using your calculator to evaluate Eq. (3-15) at selected values of t. The numerical results are slightly higher than the analytical ones. For the conditions of this example, the error in the numerical results is 1.7 percent at $t = 5$ and 1.3 percent at $t = 15$. Their accuracy can be improved by running with a smaller value of Δt. For instance, with $\Delta t = 0.25$, the error drops to 0.87 percent at $t = 5$. ////

The physical reasons for the behavior of the skydiver are apparent. In

Table 3-3. (HP-25) Fall with friction proportional to $(dy/dt)^2$

Register Contents:

0	1	2	3	4	5	6	7
—	Δt	g	α	y	dy/dt	t	—
	preloaded						

Program

Step	Code	Key Entry	X	Y	Z	T	Comments
00	00	0	0				
01	23 04	STO 4	0				4 zeroed
02	23 05	STO 5	0				5 zeroed
03	23 06	STO 6	0				6 zeroed
04	24 02	RCL 2	g				Start loop
05	24 03	RCL 3	α	g			
06	24 05	RCL 5	dy/dt	α	g		
07	15 02	g x^2	$(dy/dt)^2$	α	g		
08	61	×	$\alpha(dy/dt)^2$	g			
09	41	−	C				$C = g - \alpha(dy/dt)^2$
10	24 06	RCL 6	t	C			
11	15 71	g $x=0$	t	C			Test t for routing to put 1/2 in preliminary loop
12	13 26	GTO 26	t	C			
13	22	R↓	C				Will be in $C/2$ in prelim. loop
14	24 01	RCL 1	Δt	C			
15	61	×	$C\,\Delta t$				5 now $dy/dt_{1/2}$ in prelim. loop
16	23 51 05	STO + 5	$C\,\Delta t$				5 now $dy/dt_{3/2}$ in next loop

Step	Code	Instruction			Comment
18	24 01	RCL 1		Δt	
19	23 51 06	STO + 6		Δt	6 now t_1 in prelim. loop
					6 now t_2 in next loop
20	24 05	RCL 5	Δt	dy/dt	
21	61	x		$\Delta t\ dy/dt$	
22	23 51 04	STO + 4		$\Delta t\ dy/dt$	4 now y_1 in prelim. loop
					4 now y_2 in next loop
23	24 04	RCL 4		y	
24	14 74 (or 74)	f PAUSE (or R/S)		y	
25	13 05	GTO 05		y	To new loop
26	22	R↓	C	C	
27	02	2		2	
28	71	÷		$C/2$	
29	13 15	GTO 15		$C/2$	To continue loop

57

Table 3-3. (SR-56) Fall with friction proportional to $(dy/dt)^2$

Register Contents:

0	1	2	3	4	5	6	7	8	9
—	Δt	g	α	y	dy/dt	t	$C, C/2$	—	—

preloaded (under registers 1, 2, 3)

Program

Step	Code	Key Entry	Comments
00	00	0	0
01	33	STO	
02	04	4	4 zeroed
03	33	STO	
04	05	5	5 zeroed
05	33	STO	
06	06	6	6 zeroed
07	56	2nd CP	Test register zeroed
08	34	RCL	Start loop
09	02	2	g
10	74	−	
11	34	RCL	
12	03	3	α
13	64	×	
14	34	RCL	
15	05	5	dy/dt
16	43	x^2	$(dy/dt)^2$
17	94	=	$C = g - \alpha(dy/dt)^2$

Step	Code	Key Entry	Comments
31	35	SUM	5 now $dy/dt_{1/2}$ in prelim. loop, $dy/dt_{3/2}$ in next
32	05	5	
33	34	RCL	Δt
34	01	1	
35	35	SUM	6 now t_1 in prelim. loop, t_2 in next
36	06	6	
37	64	×	$dy/dt_{1/2}$ in prelim. loop, $dy/dt_{3/2}$ in next
38	34	RCL	
39	05	5	$\Delta t\, dy/dt_{1/2}$ in prelim. loop,
40	94	=	$\Delta t\, dy/dt_{3/2}$ in next
41	35	SUM	4 now y_1 in prelim. loop, y_2 in next
42	04	4	
43	34	RCL	

Line	Code	Key	Comment
18	33	STO	
19	07	7	7 now C
20	34	RCL	
21	06	6	t
22	37	2nd x = t	Test t for routing in preliminary loop
23	05	5	
24	00	0	
25	34	RCL	
26	01	1	
27	64	×	
28	34	RCL	
29	07	7	Δt
30	94	=	$C/2$ in prelim. loop, C in next loop / $\Delta t C/2$ in prelim. loop. $\Delta t C$ in next loop

Line	Code	Key	Comment
44	04	4	y_1 in prelim. loop, y_2 in next
45	59 (or 41)	2nd PAUSE (or R/S)	
46	59 (or 46)	2nd PAUSE (or 2nd NOP)	
47	22	GTO	
48	00	0	
49	08	8	
50	02	2	To new loop
51	12	INV	2
52	30	2nd PROD	
53	07	7	7 now $C/2$
54	22	GTO	
55	02	2	To continue loop
56	05	5	

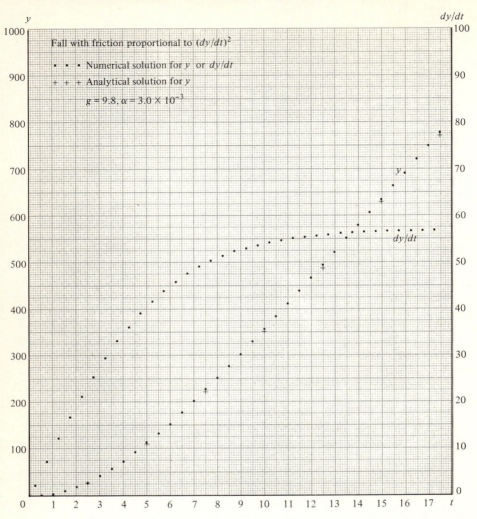

Figure 3-6. Analytical and half-increment numerical results for a body falling from rest with friction proportional to the square of its speed.

the first second or two of falling from rest, the speed is so low that the resulting air resistance is small and the only significant force in action is the gravitational force. Acceleration at the beginning of the jump is almost equal to g, and the distance traveled increases in approximate proportion to the square of the elapsed time, as with free fall. But with continued acceleration, speed builds up rapidly and the frictional force of air resistance builds up even more rapidly, since it is proportional to the square of the speed. Since this force opposes the gravitational force, the net force acting on the skydiver becomes smaller and smaller. The effect is to reduce acceleration, or rate of change of speed. The skydiver's speed stops changing when it reaches just that value for which the frictional force exactly cancels the gravitational force.

It is interesting to go through a similar analysis of the mathematical reasons for the behavior of the solution to the differential equation for the skydiver. The term d^2y/dt^2 on the left side of Eq. (3-10) is, at any value of t, a measure of the *curvature* of a curve traced out by a plot of y versus t. The reason is that $d^2y/dt^2 = d(dy/dt)/dt$; that is, it is the rate of change of the slope dy/dt of the curve. So if the slope is changing rapidly at some t and the curve, therefore, has a high curvature at that t, then d^2y/dt^2 will be large. In any region of t where d^2y/dt^2 is positive, the slope dy/dt is always becoming more positive with increasing t and the curve is *concave upward*; if d^2y/dt^2 is negative in some region of t, the curve is *concave downward* in that region.

Now note that d^2y/dt^2 in Eq. (3-10) has a positive value, g, at $t = 0$ for the case at hand, where $dy/dt = 0$ at $t = 0$. This means that dy/dt will become positive as t increases. But it can never exceed $\sqrt{g/\alpha}$, since the closer it gets to that value the closer its rate of change d^2y/dt^2 gets to zero. Therefore, d^2y/dt^2 is initially positive and subsequently decreases to zero without ever becoming negative. The result is that the curve traced out by the plot of y versus t will be concave upward for small t and subsequently loose its curvature without ever becoming concave downward. That is, it gradually approaches a straight line with slope equal to the terminal value $dy/dt = \sqrt{g/\alpha}$.

3-6. SUMMARY OF HALF-INCREMENT METHOD

There is a general prescription for applying the half-increment method to *any* second-order (containing no higher than second derivatives) ordinary (containing no partial derivatives) differential equation. You write the equation as

$$\frac{d^2y}{dt^2} = C \tag{3-16}$$

where C may involve any functions of any or all of the quantities y, t, and dy/dt, as well as any constants. Then write the three half-increment equations, with C the coefficient of $\Delta t/2$ in the first and the coefficient of Δt in the second:

$$\frac{dy}{dt_{1/2}} \simeq \frac{dy}{dt_0} + C\frac{\Delta t}{2} \tag{3-17}$$

$$\frac{dy}{dt_{i+1/2}} \simeq \frac{dy}{dt_{i-1/2}} + C\Delta t \tag{3-18}$$

$$y_{i+1} \simeq y_i + \frac{dy}{dt_{i+1/2}}\Delta t \tag{3-19}$$

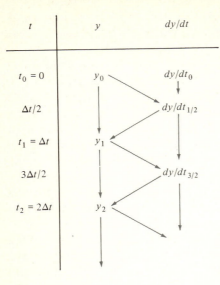

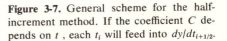

Figure 3-7. General scheme for the half-increment method. If the coefficient C depends on t, each t_i will feed into $dy/dt_{i+1/2}$.

The associated general scheme is shown in Fig. 3-7; note that there are also arrows leading down and to the right to allow for the possibility that C contains some function of y. Not shown are arrows connecting each t_i with $dy/dt_{i+1/2}$, which would be present if C contains some function of t. You can justify these equations and scheme either by deriving them by a half-increment modification of a generalized form of the derivation in Sec. 3-2 or, more simply, by noting that they immediately produce Eqs. (3-7), (3-8), and (3-9), or (3-12), (3-13), and (3-14) when C is set equal to g or $g - \alpha \, (dy/dt)^2$. Anyway, their validity is obvious. The first two equations say that the new value of the first derivative approximately equals the old value plus the rate of change of the first derivative, C, multiplied by the appropriate change in the independent variable; the third equation says the new value of the dependent variable approximately equals the old value plus the rate of change of the dependent variable, dy/dt, multiplied by the change in the independent variable. In advanced work you may encounter a C for which there is a problem in applying the method because it becomes infinite within the range where it must be evaluated. If this ever happens, try joining the numerical solution to a Green's function solution (Ref. 4) for the point where C becomes infinite.

   If the dependent and/or independent variables are other than y and t, all you have to do is rewrite these equations with the appropriate symbols. It is easy to reduce the half-increment method so that it will solve first-order, ordinary differential equations. It can also be extended to work with third- or higher-order equations, but this is not commonly required in physics. The method for second-order equations can be applied to handle sets of coupled equations of this type, as you will see in Chaps. 5 and 6.

Most of the partial differential equations arising in physics can be separated into sets of uncoupled, ordinary differential equations, which then can be solved by the half-increment method.

It is difficult to predict the accuracy of the method, but it is almost always true that: (1) the accuracy increases as you decrease the size of the increment; (2) the larger the curvature of the solution to the equation, the smaller the size of the increment you must use to achieve a certain accuracy; (3) the accuracy decreases when the C coefficient involves a first derivative. You can understand the third comment by inspecting Eq. (3-18) and Fig. 3-7 and noting that the dependent and independent variables are both evaluated in C at the middle of the interval where C is used, whereas if the first derivative is present in C it is evaluated at the beginning of the interval. More complicated numerical methods overcome this source of inaccuracy, but the limitations of programmable pocket calculators generally preclude their use of these methods. Fortunately, it is frequently true that where the first derivative is large its rate of change is small so that the value of C is insensitive to where the first derivative is evaluated. The case you studied in Sec. 3-5 is an example. In very infrequently occurring circumstances, the numerical solution to a differential equation is unstable in that a point is reached where further decrease in the size of the increment leads to decrease in the accuracy. If you ever come across such an example, consult an applied mathematician.

EXERCISES

3-1. Run the full-increment free-fall program with a value of Δt smaller by a factor of $\frac{1}{2}$ than that used in the example. Compare the results with the analytical solution for free fall plotted in the example.

3-2. Do the same for the half-increment program. Explain why for all values of Δt the half-increment solution for free fall is in exact agreement with the analytical solution. Why would this not be true when the half-increment procedure is applied to problems other than free fall?

3-3. Do the same for the half-increment program for fall with friction proportional to the square of the speed. Compare the results with your discussion of 3-2.

3-4. The gravitational acceleration of an object falling freely near the surface of the moon is 1.6 m/s². An astronaut accidentally falls from the ladder leading to the door of the moon lander at a height of 5m. Use the half-increment program to determine the time until impact on the surface. Also determine the speed at impact by recalling the contents of the register that contains dy/dt.

3-5. Run the fall with friction proportional to the square-of-speed program with $\alpha = 3.0 \times 10^{-2}$. Compare the results with the example.

3-6. Change the fall-with-friction program to make the frictional force proportional to the first power of speed by simply deleting the step in which the speed is squared. Then find a value for α which will lead to the same terminal speed as in the squared-speed program example. Run your modified program with this α, plot y and dy/dt, and compare with the plotted example. Discuss the difference.

3-7. Do the same, but with the program modified so that friction is proportional to the cube of the speed. This is of practical importance in certain cases, but there is no analytical solution to the differential equation. *Note:* If your calculator objects to evaluating y^x for $y = 0$, fool it by setting the initial value of dy/dt equal to some very small number.

3-8. Study the details of the programs in Tables 3-1 and 3-3 and explain precisely what happens in each step.

3-9. Write a modification of the fall-with-friction program that will cause $j - 1$ successive values of y to be calculated without being displayed. By displaying only every jth value, it is convenient to improve accuracy by reducing the value of Δt, without increasing the work required to plot results. Try it with $j = 10$, $\Delta t = 0.05$.

3-10. Write a half-increment program to solve the differential equation for exponential growth, or decay,

$$\frac{dN}{dt} = +NR \qquad \text{or} -NR$$

Then study the properties of the solutions for various values of the growth, or decay, rate R. Explain which electrical circuits are governed by the differential equation for decay and how the circuit parameters are related to R. Do the same for exponential growth of bacteria.

REFERENCES

1. Bueche, Frederick J.: *Introduction to Physics for Scientists and Engineers,* 2d ed., McGraw-Hill Book Company, New York, 1975, p. 54ff.
   Halliday, David, and Robert Resnick: *Fundamentals of Physics,* John Wiley & Sons, Inc., New York, 1974, p. 63.
   Sears, Francis W., and Mark W. Zemansky: *University Physics,* 4th ed., Addison-Wesley Publishing Company, Inc., Reading, Mass., 1970, pp. 15–16, 57.
   Weidner, Richard, and Robert Sells: *Elementary Classical Physics,* Allyn and Bacon, Inc., Boston, 1973, p. 106.
2. Ford, Kenneth: *Classical and Modern Physics,* Xerox, Lexington, Mass., 1973, p. 513.
3. Ford, Kenneth: *Classical and Modern Physics,* Xerox, Lexington, Mass., 1973, p. 515.
4. Dennery, Philippe, and André Krzywicki: *Mathematics for Physicists,* Harper & Row, Publishers, Inc., New York, 1967, p. 273.

FOUR
OSCILLATORS

4-1. INTRODUCTION

Now that a method for solving differential equations numerically has been developed and its use with programmable pocket calculators has been explained, the thrust of most of the remainder of this book will be to show how it can help you study the behavior of many different physical systems.

Some of the most important of these are systems involving oscillations. The oscillations can be in mechanical objects, like masses connected to springs or pendulums. In fact these are examples you will come across in the next two chapters, but other mechanical examples could just as well have been used—the timing mechanism in a watch, the string of a guitar, atoms in a heated solid, or in a gas transmitting a sound wave. Furthermore, very important nonmechanical systems involve oscillations, e.g., lightwaves, radiowaves, and electrical circuits. Oscillatory motion is strongly emphasized in most elementary physics books because it is so wide spread in nature, despite the fact that solving the differential equations that arise by analytical methods is not feasible at the elementary level for many interesting cases, and cannot be done at all in some cases. Such difficulties are not present, however, if numerical methods are used.

4-2. HARMONIC OSCILLATIONS

Figure 4-1 shows the first oscillating system you will consider. An object of mass m is at one end of a horizontal spring whose other end is fixed. Since gravity does not act in the horizontal direction, the only horizontal force exerted on the object arises from the extension or compression of the spring relative to its normal length. The coordinate x measures the displacement of the object from its position when the spring has its normal length and, therefore, also the extension or compression of the spring.

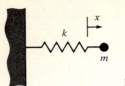

Figure 4-1. An example of a harmonic oscillator.

The quantity k is a measure of the stiffness of the spring and is the proportionality constant in the relation

$$F = -kx \qquad (4\text{-}1)$$

between the extension or compression of the spring and the force F which it exerts on the object. This experimental relation is called *Hooke's law* and applies to anything capable of flexing, providing the flexure does not approach its "elastic limit." (See Ref. 1.) Note the minus sign, which properly makes F negative if x is positive, and vice versa, so that the spring pulls on the object if extended and pushes on it if compressed.

Newton's law of motion

$$\frac{d^2x}{dt^2} = \frac{F}{m}$$

contains complete predictions about the possible behavior of the system. To obtain them, the first step is to use Eq. (4-1) to evaluate F, which yields

$$\frac{d^2x}{dt^2} = -\frac{k}{m}x \qquad (4\text{-}2)$$

or

$$\frac{d^2x}{dt^2} = -\alpha x \qquad (4\text{-}3)$$

where

$$\alpha = \frac{k}{m} \qquad (4\text{-}4)$$

The second step is to solve this differential equation.

The prescription of Sec. 3-6 makes the job almost automatic. Comparing the differential equation with Eq. (3-16), the coefficient C is seen to have the value $C = -\alpha x$. Using this in Eqs. (3-17), (3-18), and (3-19), and also changing the dependent variable in them from y to x, the formulas you need are obtained immediately:

$$\frac{dx}{dt}_{1/2} \simeq \frac{dx}{dt_0} + C\frac{\Delta t}{2} \qquad C = -\alpha x_0 \qquad (4\text{-}5)$$

$$\frac{dx}{dt}_{i+1/2} \simeq \frac{dx}{dt}_{i-1/2} + C\Delta t \qquad C = -\alpha x_i \qquad (4\text{-}6)$$

$$x_{i+1} \simeq x_i + \frac{dx}{dt}_{i+1/2}\Delta t \qquad (4\text{-}7)$$

The scheme is just the one shown in Fig. 3-7, except that y should be read as x.

Tables 4-1 (SR-56) and (HP-25) contain the programs. Instructions for using them follow.

For the SR-56

Clear by keying **2nd CP,** key **LRN,** then enter the indicated key strokes through step 08. Step 09 is a no-operation instruction which simply makes the calculator go to the next step. It is entered to leave room for the operation of computing $\sin x$, which will be used in Sec. 4-3 to treat the case of a pendulum. Next key in steps 10 through 38. If you are only planning to watch the display, key **2nd pause** in steps 39 and 40; if you are planning to graph the results, key **R/S** in the first and **2nd NOP** in the second. Then key the remainder of the program and return to the execute mode by keying **LRN.** After setting the calculator to step 00 by keying **RST,** load the values you choose for initial x, initial dx/dt, Δt, and α by keying the first followed by **STO 0,** the second by **STO 1,** the third by **STO 2,** and the fourth by **STO 3.** Then press **R/S** to start. If step 39 is **R/S,** the calculator will stop there waiting for you to press **R/S** for the next loop. To see the value of t when stopped by a program **R/S,** key **RCL 4.** To see t in the pause mode, key **R/S** after pausing and then key **2nd EXC 4,** inspect t, then key **2nd EXC 4** again. The program always takes the initial value of t to be zero.

For the HP-25

Switch to **PRGM,** clear by keying **f PRGM,** and enter the indicated key strokes through step 04. Step 05 is a no-operation instruction which simply makes the calculator go to the next step. It is entered to leave room for the operation of computing $\sin x$, which will be used in Sec. 4-3 to treat the case of a pendulum. Key in steps 06 through 19 next. If you are only planning to watch the display, key **f PAUSE** in step 20; if you are planning to graph the results, key **R/S.** Then key the remainder of the program and switch back to **RUN.** After setting the calculator to step 00 by keying **f PRGM,** load the values you choose for initial x, initial dx/dt, Δt, and α by keying the first followed by **STO 0,** the second by **STO 1,** the third by **STO 2,** and the fourth by **STO 3.** Then press **R/S** to start. If step 20 is **R/S,** the calculator will stop there waiting for you to press **R/S** for the next loop. To see the value of t when stopped by a program **R/S,** key **RCL 4.** To see t in the pause mode, key **R/S** while pausing and then key **RCL 4.** The program always takes the initial value of t to be zero.

Examples

initial $x = 0.25$, initial $dx/dt = 0$, $\Delta t = 0.2$, $\alpha = 1$

initial $x = 1.50$, initial $dx/dt = 0$, $\Delta t = 0.2$, $\alpha = 1$

Table 4-1. (SR-56) Oscillations

Register Contents:

0	1	2	3	4	5	6	7	8	9
x	dx/dt	Δt	α	t	$C,\ C/2$	—	—	—	—

preloaded

run in radian mode for pendulum

Program

Step	Code	Key Entry	Comments
00	00	**0**	0
01	33	**STO**	
02	04	**4**	4 zeroed
03	56	**2nd CP**	Test register zeroed
04	34	**RCL**	Start loop
05	03	**3**	α
06	64	**x**	
07	34	**RCL**	
08	00	**0**	x
09	46	**2nd NOP**	Allows room for pendulum program
10	94	**=**	αx
11	93	**+/-**	$C = -\alpha x$
12	33	**STO**	
13	05	**5**	5 now C
14	34	**RCL**	
15	04	**4**	t
16	37	**2nd x = t**	Test t for routing in preliminary loop
17	04	**4**	

Step	Code	Key Entry	Comments
27	34	**RCL**	Δt
28	02	**2**	
29	35	**SUM**	
30	04	**4**	4 now t_1 in prelim. loop, t_2 in next
31	64	**x**	$dx/dt_{1/2}$ in prelim. loop, $dx/dt_{3/2}$ in next
32	34	**RCL**	
33	01	**1**	$\Delta t\, dx/dt_{1/2}$ in prelim. loop, $\Delta t\, dx/dt_{3/2}$ in next
34	94	**=**	
35	35	**SUM**	0 now x_1 in prelim. loop, x_2 in next
36	00	**0**	
37	34	**RCL**	x_1 in prelim. loop, x_2 in next
38	00	**0**	
39	59 (or 41)	**2nd PAUSE** (or R/S)	
40	59 (or 46)	**2nd PAUSE** (or 2nd NOP)	

Line	Code	Key	Comment
18	04	4	
19	34	RCL	
20	02	2	Δt
21	64	×	
22	34	RCL	$C/2$ in prelim. loop, C in next loop
23	05	5	
24	94	=	$\Delta tC/2$ in prelim. loop, ΔtC in next loop
25	35	SUM	1 now $dx/dt_{1/2}$ in prelim. loop, $dx/dt_{3/2}$ in next
26	01	1	

Line	Code	Key	Comment
41	22	GTO	To new loop
42	00	0	
43	04	4	2
44	02	2	
45	12	INV	
46	30	2nd PROD	5 now $C/2$
47	05	5	
48	22	GTO	
49	01	1	To continue loop
50	09	9	

69

Table 4-1. (HP-25) Oscillations

Register Contents:

0	1	2	3	4	5	6	7
x	dx/dt	Δt	α	t	—	—	—

preloaded

run in radian mode for pendulum

Program

Step	Code	Key Entry	X	Y	Z	T	Comments
00	00	0	0				
01			0				4 zeroed
02	23 04	STO 4					Start loop
03	24 03	RCL 3	α				
04	24 00	RCL 0	x	α			Allows room for change to be made in pendulum program
05	15 74	g NOP	x	α			
06	61	x	$-C$				$C = -\alpha x$
07	24 04	RCL 4	t	$-C$			
08	15 71	g x = 0	t	$-C$			Test t for routing to put 1/2 in preliminary loop
09	13 22	GTO 22	t				
10	22	R↓	$-C$	$-C$			
11	24 02	RCL 2	Δt	$-C$			Will be $-C/2$ in prelim. loop
12	61	x	$-C\Delta t$				1 now $dx/dt_{1/2}$ in prelim. loop
13	23 41 01	STO – 1	$-C\Delta t$				1 now $dx/dt_{3/2}$ in next loop

	Code		Instruction		Stack		
14		24 02	RCL 2		Δt		4 now t_1 in prelim. loop
15	23	51 04	STO + 4		Δt		4 now t_2 in next loop
16		24 01	RCL 1	Δt	dx/dt		
17		61	x		$\Delta t\ dx/dt$		0 now x_1 in prelim. loop
18	23	51 00	STO + 0		$\Delta t\ dx/dt$		0 now x_2 in next loop
19		24 00	RCL 0		x		
20		14 74	f PAUSE		x		
		(or 74)	(or R/S)				
21		13 03	GTO 03		x		To new loop
22		22	R↓		$-C$		
23		02	2		2		
24		71	÷	$-C$	$-C/2$		
25		13 11	GTO 11		$-C/2$		To continue loop

The results for both are plotted in Fig. 4-2. In both, if $\alpha = 1$ means $\alpha = 1$ N/m·kg (Newtons per meter-kilogram), then x will be in meters and t in seconds. Of course, you can use any system of units you want for the input numbers, providing it is a consistent system. The calculator will output values in the appropriate units of that system. As the physics to be learned from all this does not depend on the units, they will usually be ignored. ////

In the first example the oscillator was displaced to $x = 0.25$ and then let go from rest. You can see how it slowly starts to move to $x = 0$, picks up speed, passes through $x = 0$, slows down as it approaches $x = -0.25$, turns around there, repeats its motion in the opposite direction until it again gets to $x = 0.25$, and then starts the next identical *cycle*. The points look like they would fall accurately on a sinusoidal curve (a cosine), and indeed they would. As sinusoidal functions are sometimes called *har-*

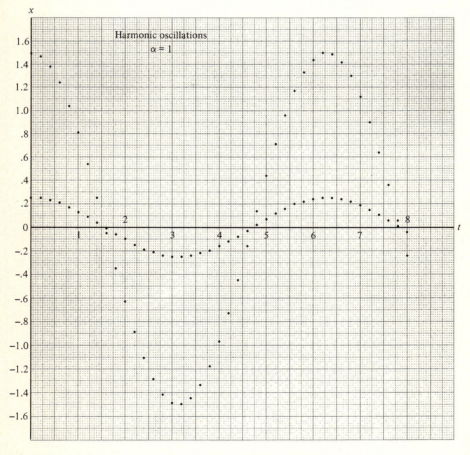

Figure 4-2. Harmonic oscillations of small and large amplitudes.

monic functions, an oscillator satisfying the differential equation (4-3) is sometimes called a *harmonic oscillator.*

The *period T* of the oscillation is the time required to complete one full cycle, and the *frequency ν* is $1/T$. You can measure T from a plot of x versus t or, less accurately, without plotting by stopping the calculator when you see x change sign at the beginning of a cycle, recording the contents of the t register, then doing it again at the beginning of the next cycle. Careful inspection of Fig. 4-2 will show you that for the $α = 1$ oscillator, $T \simeq 6.28$ and $ν \simeq 1/6.28$. Do these numbers remind you of 2π and $1/2\pi$? They are. Can you think of how to run the program to verify this with more accuracy, but without plotting?

In the second example the oscillator was initially displaced to $x = 1.50$ and released with no initial speed. It then proceeded to execute oscillations which look very much like those it made in the first example, except with all the values of x scaled up in magnitude by a factor of $1.50/0.25$. If you look carefully at the plots, or the numbers you can record by running the programs yourself, you will find that this is the case. The maximum magnitude of x in an oscillation is called the *amplitude.* This *scaling property* means that the characteristics of the oscillations, such as its period or frequency, do not depend on the amplitude. The property reflects the fact that the differential equation governing the oscillations, Eq. (4-3), is *linear* in x. That is, each term in it is proportional to x to the first power. Thus if a certain dependence of x on t solves the equation, then the same dependence, but with x scaled up or down by any constant factor you wish, will also be a solution.

There are two interesting questions to ask: (1) Why does the oscillator oscillate?, and (2) Why does the differential equation associated with the oscillator have oscillatory solutions?

The first requires a physical explanation. If the object at the end of the spring is pulled from its normal position and then released from rest, the force exerted on it by the spring is free to act. It will pull on the object in the direction toward the normal position because the spring is stretched. This accelerates the object toward that position, and so it picks up speed while moving in that direction. At the instant it passes through the normal position, there is no force acting on it because the spring is at its normal length. But the object must keep moving because it has mass and, therefore, momentum which cannot change if no force is applied. As it continues the body compresses the spring, so the spring exerts a force on it acting back toward the normal position of the body. This slows it until it comes momentarily to rest, having completed the first half cycle of an oscillation. The next half cycle is just the reverse of the first, and all subsequent full cycles are like those that precede.

The second question is answered by analyzing the relation imposed by the differential equation (4-1) between the quantity x and its second derivative d^2x/dt^2. As explained in Chap. 3, d^2x/dt^2 is a measure of the curvature of a plot of x versus t. If d^2x/dt^2 is positive the curvature is concave upward, and if it is negative the curvature is concave downward. Since the quantity α is positive, the differential equation $d^2x/dt^2 = -\alpha x$ says that the sign of the second derivative of x is always opposite to the sign of x itself. Thus if in some region of t the curve traced out by plotting x versus t lies above the t axis, then x is positive, d^2x/dt^2 is negative, and the curve is concave downward. If the curve lies below the t axis it is concave upward. Simply put, the curve is under all circumstances concave toward the t axis. This means that the curve must oscillate about the axis, so x is an oscillatory function of t.

4-3. LARGE OSCILLATIONS OF A PENDULUM

Figure 4-3 represents a pendulum consisting of a body of mass m at the end of a cord of length l and negligible mass. At the instant shown, the cord makes an angle θ with respect to the vertical. The two forces acting on the body are the tension T in the cord, which acts along its direction, and the gravitational force, or weight mg, which acts in the downward direction. In Fig. 4-4 the weight is resolved into two perpendicular components. One is in the direction opposite to the tension and the other is directed along the tangent to the path of the body at the instant illustrated. The latter, which is of magnitude $mg \sin \theta$, is the one of interest because it is the restoring force that leads to the oscillatory motion of the pendulum.

To study the motion, you may apply Newton's law to the tangential direction at the instant illustrated by writing $x = l\theta$, where x is distance traveled by the body along its circular path, measured from its vertical position, and θ is measured in radians. Then

$$\frac{d^2x}{dt^2} = \frac{d^2(l\theta)}{dt^2} = l\frac{d^2\theta}{dt^2}$$

since l is a constant. If the last step escapes you, look at the definition of a derivative in Chap. 1 and apply it to the derivative of the product of a con-

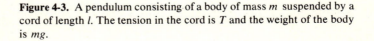

Figure 4-3. A pendulum consisting of a body of mass m suspended by a cord of length l. The tension in the cord is T and the weight of the body is mg.

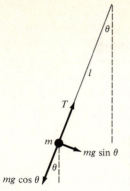

Figure 4-4. The weight is resolved into components tangential and perpendicular to the path of the body at the instant shown. The tangential component $mg \sin \theta$ provides the restoring force that makes the pendulum oscillate.

stant times a variable. With this relation substituted into Newton's law, you have

$$l \frac{d^2\theta}{dt^2} = \frac{-mg \sin \theta}{m}$$

or

$$\frac{d^2\theta}{dt^2} = -\alpha \sin \theta \qquad (4\text{-}8)$$

where

$$\alpha = \frac{g}{l} \qquad (4\text{-}9)$$

The minus sign reflects the fact that the sign of the restoring force is always opposite to the sign of the angular coordinate θ, as inspection of the figures will demonstrate.

Note the similarity of Eqs. (4-8) and (4-3). Except for the trivial differences that one involves θ and the other x, and that the meaning of α is not the same, the two equations would be identical if it were true that $\sin \theta = \theta$. Of course, that is not generally true. But it is approximately true if the angle θ, expressed in radians, is small compared to 1. (If you are unfamiliar with that fact, set your calculator to the radian mode by keying **2nd RAD** on an SR-56 or by keying **g RAD** on an HP-25, and try it for $\theta = 0.5, 0.05,$ and 0.005.)

The standard approach of elementary physics textbook treatments of the pendulum at this point is to use the $\sin \theta = \theta$ approximation so that the pendulum differential equation becomes essentially identical to the harmonic-oscillator differential equation. The reason is that the change in the form of the differential equation, caused by the presence of the sine of the dependent variable instead of the dependent variable itself, converts it from one that can be solved analytically without too much difficulty to one that can be solved only by numerical methods.

But it is not necessary for you to make the approximation, and therefore preclude yourself from investigating the large oscillations of a pendulum, because it is extremely easy to modify the program in Tables 4-1 (SR-56) or (HP-25) so that it will solve Eq. (4-8).

For the SR-56

Assuming that your calculator is still programmed to run the harmonic oscillator, while in the execute mode key **GTO 0 9**, key **LRN** key **sin**, key **LRN**, and key **RST** to reset to step 00. You have instructed the calculator to take the sine of the dependent variable in step 09, so it will now solve Eq. (4-8). Since the dependent variable is θ, measured in radians, *be sure* to key **2nd RAD** to put it in the radian mode before running. Then operate it just as before.

For the HP-25

Assuming that your calculator is still programmed to run the harmonic oscillator, while switched to **RUN** key **GTO 0 4**, switch to **PRGM**, key **f sin**, switch back to **RUN**, and key **f PRGM** to reset to step 00. You have instructed the calculator to take the sine of the dependent variable at step 05, so it will now solve Eq. (4-8). Since the dependent variable is θ, measured in radians, *be sure* to key **g RAD** to put it in the radian mode before running. Then operate it just as before.

Examples

initial $\theta = 0.25$, initial $d\theta/dt = 0$, $\Delta t = 0.2$, $\alpha = 1$
initial $\theta = 1.50$, initial $d\theta/dt = 0$, $\Delta t = 0.2$, $\alpha = 1$

For both examples the results are plotted in Fig. 4-5. Compare the oscillation in θ of amplitude 0.25 with the harmonic oscillation in x of the same amplitude plotted in Fig. 4-2. It is clear that for an amplitude of 0.25 rad = 14°, the motion of a pendulum is a very excellent approximation to harmonic oscillation.

Not so for pendulum motion with an amplitude of 1.50 rad = 86°, as the plot of the second example shows. The points for this plot do not trace out a harmonic (or sinusoidal) curve, although they certainly fall on an oscillatory curve. Note that the period is lengthened, or frequency reduced, for the large-amplitude oscillation. Can you use the force diagram in Fig. 4-4 to explain from a physical point of view the origin of this behavior?////

The plot for large-amplitude oscillations of the pendulum does not look like the one for small-amplitude oscillations with all the values of θ multiplied by some scale factor. The shape changes, although you may regard the change as rather subtle. But the change in the period is certainly apparent; so it is equally apparent that the scaling property discussed in connection with the harmonic oscillator does not apply to the pendulum.

The reason for this involves the fact that the differential equation for a pendulum, (4-8), is not linear in θ. If you reread the end of Sec. 4-2, you should be able to explain it to yourself from a mathematical point of view.

The *nonlinearity* of Eq. (4-8) is also the reason why it can be solved only by numerical methods. As a rule of thumb, you can expect this to be

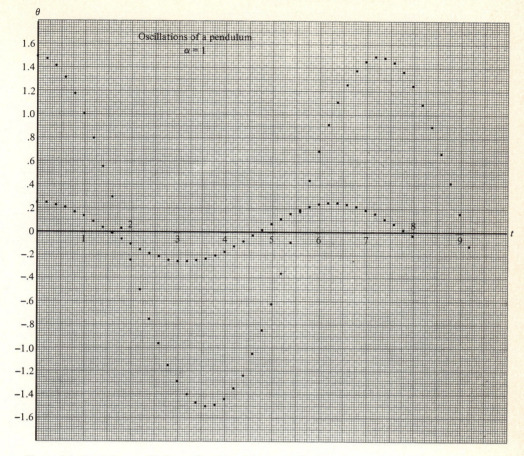

Figure 4-5. Pendulum oscillations of small and large amplitudes.

the case for most nonlinear differential equations. But there are some lucky exceptions such as Eq. (3-10).

4-4. DAMPED, DRIVEN OSCILLATIONS

Next you will reconsider the harmonic oscillator shown in Fig. 4-1, allowing for two additional possibilities: (1) there may be appreciable friction (e.g., the body at the end of the spring is moving through maple syrup); (2) a periodically varying force may be applied to the body by some external agency (e.g., you are pushing on it rhythmically). Since most oscillating bodies are not large ones moving rapidly, it is reasonable to assume that the magnitude of the frictional force they experience is proportional to the first power of their speed (see Sec. 3-5.) Another reason for making this assumption is that with a frictional *damping* force proportional to dx/dt, the differential equation has exactly the same mathematical form as one arising in certain aspects of electromagnetism. So you will be able use it and what you learn about its properties if you want to study that subject (see Ref. 2). As for the applied *driving* force, it is convenient and yet sufficiently illustrative to take it to be proportional to $\sin \omega t$, where ω is its angular frequency in radians per second. That is, the oscillator is driven with a sinusoidally varying force of ordinary frequency $\nu = \omega/2\pi$ Hz (cycles per second).

Equation (4-2) for the undamped, undriven oscillator, which came directly from Newton's law, was

$$\frac{d^2x}{dt^2} = \frac{F}{m} = \frac{-kx}{m}$$

Adding to the *restoring* force, $-kx$, terms representing damping and driving forces gives

$$\frac{d^2x}{dt^2} = \frac{-kx - f\dfrac{dx}{dt} + a \sin \omega t}{m}$$

where f specifies the damping force for unit speed and a specifies the amplitude of the driving force. The damping force term is negative, just as in Eq. 3-10, because friction always tends to oppose the direction of motion specified by the sign of dx/dt. Introducing

$$\alpha = \frac{k}{m} \qquad \beta = \frac{f}{m} \qquad \gamma = \frac{a}{m} \tag{4-10}$$

allows the equation to be written as

$$\frac{d^2x}{dt^2} = -\alpha x - \beta \frac{dx}{dt} + \gamma \sin \omega t$$

or
$$\frac{d^2x}{dt^2} = C \qquad C = -\alpha x - \beta \frac{dx}{dt} + \gamma \sin \omega t \tag{4-11}$$

Now that the form of the coefficient C has been identified, the set of basic equations for the numerical method analogous to Eqs. (4-5), (4-6), and (4-7) can be written immediately. But it is hardly necessary to do so to make the additions required in the harmonic-oscillator program to allow it to handle the present case. As you can see from inspecting Tables 4-2 (SR-56) or (HP-25), the only change required is the somewhat more complicated routine used to generate the new form of C.

Both versions of the program are keyed in and run just as for the oscillator program, except for the following changes.

For the SR-56

The "no operation" instruction is absent, as a blank step will not be needed for a subsequent modification (unless you would later like to study the damped, driven pendulum—if so, put it in step 26 and then increase all the remaining step numbers by 1). The parameters β, γ, and ω must be preloaded in registers 4, 5, and 6. This is done in just the same way that initial x, initial dx/dt, Δt, and α are inserted in registers 0, 1, 2, and 3. If you want to recall t, remember that here it is stored in register 7. *Above all* remember to put the calculator in the radian mode, before running a driven oscillator, by keying **2nd RAD** when in the execute mode. To see the driving-force term $\sin \omega t$ when the calculator has been stopped by a program **R/S**, push **SST** eleven times while remaining in the execute mode. You will now be seeing the results of step 10. Then restart by pushing **R/S**.

For the HP-25
The "no-operation" instruction is absent, as a blank step will not be needed for a subsequent modification (unless you would later like to study the damped, driven pendulum—if so, put it in step 05 and then increase all the remaining step numbers by 1). The parameters β, γ, and ω must be preloaded in registers 4, 5, and 6. This is done in just the same way that initial x, initial dx/dt, Δt, and α are inserted in registers 0, 1, 2, and 3. If you want to recall t, remember that here it is stored in register 7. *Above all* remember to put the calculator in the radian mode, before running a driven oscillator, by keying **g RAD** when switched to **RUN**. To see the driving-force term $\sin \omega t$ when the calculator is stopped and displaying x, push **SST** until you get to step 12. Then restart by pushing **R/S**.

Examples
initial $x = 1.5$, initial $dx/dt = 0$, $\Delta t = 0.2$, $\alpha = 1$, $\beta = 0.5$
initial $x = 1.5$, initial $dx/dt = 0$, $\Delta t = 0.2$, $\alpha = 1$, $\beta = 2.5$

In both cases the oscillator has the same spring-stiffness-to-mass ratio α as in the examples of an undamped oscillator plotted in Fig. 4-2, and also the same initial conditions as used for the larger-amplitude oscillation of that figure. Both cases are undriven. The first, for a damping-constant-to-mass ratio $\beta = 0.5$, is said to be *lightly damped;* the second, for $\beta = 2.5$, is said to be *heavily damped*. You can see from the results shown in Figs. 4-6 and 4-7 why this terminology is appropriate. Note that the period, or frequency, for the lightly damped oscillator appears to have the same value as in the case of the undamped oscillator with the same value of α (but a careful comparison shows the period is slightly longer). No period can be defined for the heavily damped oscillator.   ////

Can you devise a physical explanation for the behavior of the damped oscillator, and also a mathematical explanation for the behavior of the solutions to its differential equation, which combines features of the explanations given in Sec. 4-2 and Sec. 3-5?

Table 4-2. (SR-56) Damped, driven oscillations

Register Contents:

0	1	2	3	4	5	6	7	8	9
x	dx/dt	Δt	α	β	γ	ω	t	C, C/2	—

preloaded
run in radian mode if driven

Program

Step	Code	Key Entry	Comments
00	00	0	0
01	33	STO	
02	07	7	7 zeroed
03	56	2nd CP	Test register zeroed
04	34	RCL	Start loop
05	06	6	ω
06	64	x	
07	34	RCL	
08	07	7	t
09	94	=	ωt
10	23	sin	sin ωt
11	64	x	
12	34	RCL	
13	05	5	γ
14	74	−	
15	34	RCL	
16	04	4	β
17	64	x	
18	34	RCL	
19	01	1	dx/dt

Step	Code	Key Entry	Comments
37	34	RCL	$C/2$ in prelim. loop, C in next loop
38	08	8	
39	94	=	$\Delta t C/2$ in prelim. loop, $\Delta t C$ in next loop
40	35	SUM	1 now $dx/dt_{1/2}$ in prelim. loop,
41	01	1	$dx/dt_{3/2}$ in next
42	34	RCL	
43	02	2	Δt
44	35	SUM	
45	07	7	7 now t_1 in prelim. loop, t_2 in next
46	64	x	
47	34	RCL	
48	01	1	$dx/dt_{1/2}$ in prelim. loop, $dx/dt_{3/2}$ in next
49	94	=	$\Delta t\, dx/dt_{1/2}$ in prelim. loop,
50	35	SUM	$\Delta t\, dx/dt_{3/2}$ in next

Step	Code	Key	Comment
20	74		
21	34	RCL	
22	03	3	α
23	64	×	
24	34	RCL	
25	00	0	x
26	94	=	$C = \gamma \sin \omega t - \beta\, dx/dt - \alpha x$
27	33	STO	
28	08	8	8 now C
29	34	RCL	
30	07	7	t
31	37	2nd $x = t$	Test t for routing in preliminary loop
32	05	5	
33	09	9	
34	34	RCL	
35	02	2	
36	64	×	Δt

Step	Code	Key	Comment
51	00	0	0 now x_1 in prelim. loop, x_2 in next
52	34	RCL	
53	00	0	
54	59 (or 41)	2nd PAUSE (or R/S)	x_1 in prelim. loop, x_2 in next
55	59 (or 46)	2nd PAUSE or 2nd NOP	
56	22	GTO	
57	00	0	
58	04	4	To new loop
59	02	2	2
60	12	INV	
61	30	2nd PROD	
62	08	8	8 now $C/2$
63	22	GTO	
64	03	3	
65	04	4	To continue loop

Table 4-2. (HP-25) Damped, driven oscillations

Register Contents:

0	1	2	3	4	5	6	7
x	dx/dt	Δt	α	β	γ	ω	t

preloaded
run in radian mode if driven

Program

Step	Code	Key Entry	X	Y	Z	T	Comments
00	00	0	0				
01	23 07	STO 7	0				7 zeroed
02	24 03	RCL 3	α				Start loop
03	24 00	RCL 0	x	α			
04	61	x	αx				
05	24 04	RCL 4	β	αx			
06	24 01	RCL 1	dx/dt	β	αx		
07	61	x	$\beta dx/dt$	αx			
08	24 06	RCL 6	ω	$\beta dx/dt$	αx		
09	24 07	RCL 7	t	ω	$\beta dx/dt$	αx	
10	61	x	ωt	$\beta dx/dt$	αx	αx	
11	14 04	f sin	$\sin \omega t$	$\beta dx/dt$	αx	αx	
12	24 05	RCL 5	γ	$\sin \omega t$	$\beta dx/dt$		
13	61	x	$(\sin \omega t)\,\gamma$	$\beta dx/dt$	αx		
14	41	−	$\beta dx/dt - \gamma \sin \omega t$	αx			
15	51	+	$-C$	αx			$C = -(\alpha x + \beta dx/dt - \gamma \sin \omega t)$
16	24 07	RCL 7	t	$-C$			
17	15 71	g x = 0	t	$-C$			Test t for routing to put 1/2
18	13 32	GTO 32	t	$-C$			in preliminary loop

Step	Code	Instruction	Stack		Comment
20	22	R↓		$-C$	Will be $-C/2$ in prelim. loop
21	24 02	RCL 2		Δt	1 now $dx/dt_{1/2}$ in prelim. loop
22	61	x		$-C\Delta t$	1 now $dx/dt_{3/2}$ in next loop
23	23 41 01	STO − 1	$-C$	$-C\Delta t$	
24	24 02	RCL 2		Δt	7 now t_1 in prelim. loop
25	23 51 07	STO + 7		Δt	7 now t_2 in next loop
26	24 01	RCL 1		dx/dt	0 now x_1 in prelim. loop
27	61	x		$\Delta\, dx/dt$	
28	23 51 00	STO + 0	Δt	$\Delta\, dx/dt$	0 now x_2 in next loop
29	24 00	RCL 0		x	
30	14 74 (or 74)	f PAUSE (or R/S)		x	
31	13 03	GTO 03		x	To new loop
32	22	R↓		$-C$	
33	02	2		2	
34	71	÷	$-C$	$-C/2$	
35	13 21	GTO 21		$-C/2$	To continue loop

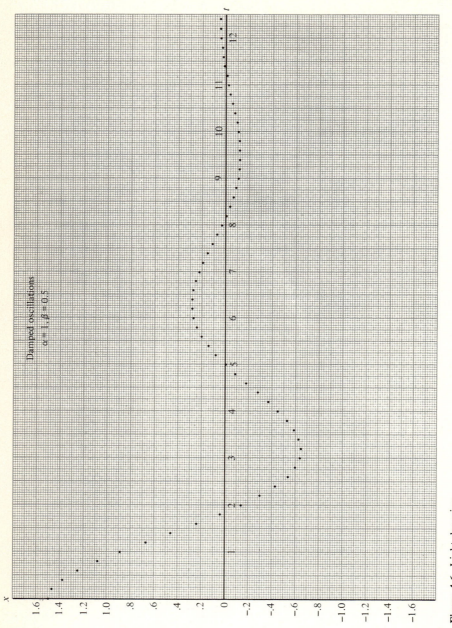

Damped oscillations

$\alpha = 1, \beta = 0.5$

Figure 4-6. Light damping.

84

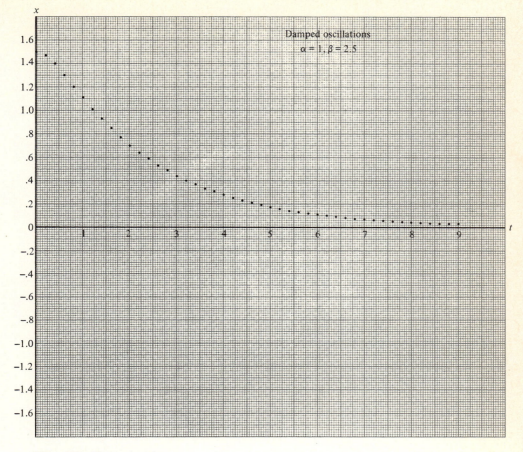

Figure 4-7. Heavy damping.

Examples

initial x = initial $dx/dt = 0$, $\Delta t = 0.8$, $\alpha = 1$, $\beta = 0.5$, $\gamma = 1$, $\omega = 0.20$

initial x = initial $dx/dt = 0$, $\Delta t = 0.2$, $\alpha = 1$, $\beta = 0.5$, $\gamma = 1$, $\omega = 0.95$

initial x = initial $dx/dt = 0$, $\Delta t = 0.2$, $\alpha = 1$, $\beta = 0.5$, $\gamma = 1$, $\omega = 2.00$

In all three examples the oscillator starts at rest from the configuration it has when the spring is at its normal length. But it does not stay there since it is driven by an applied force of unit amplitude-to-mass ratio γ which starts from zero at the initial instant as a positive-going sinewave. For all three, the stiffness-to-mass ratio α and damping-to-mass ratio β are as they were in the lightly damped oscillator plotted in Fig. 4-6, and α is the same as in the undamped oscillator plotted in Fig. 4-2.

Recall that the frequency of free oscillations for the undamped oscilla-

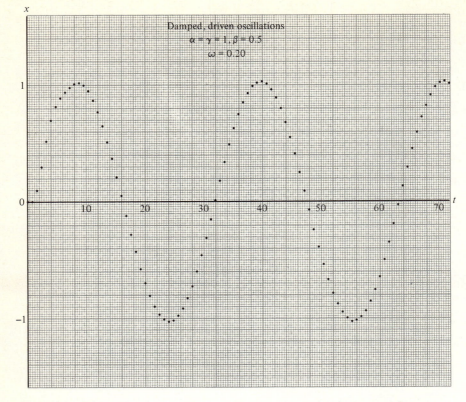

Figure 4-8. A damped oscillator driven at a frequency lower than its resonant frequency.

tor with this α, which here will be written ν_0, has the value $\nu_0 = 1/2\pi$ corresponding to an angular frequency of free oscillation $\omega_0 = 2\pi\nu_0 = 1$. Recall also that the lightly damped oscillator of this β has approximately the same oscillation frequency. In the first example, the driving frequency $\omega = 0.20$ is considerably smaller than ω_0. It is plotted in Fig. 4-8. Although the first part of the first half cycle of oscillation is noticably nonsinusoidal, the oscillator has already well adjusted its motion to the behavior imposed by the driving force by the end of the second half cycle. That is, after the first full oscillation, the system has reached a *steady state* where each cycle exactly repeats the previous cycle. You can see by inspection that the steady-state amplitude has a value slightly larger than 1.0. You should also look carefully enough to see that in steady state the oscillation frequency equals the driving frequency.

The second example is plotted in Fig. 4-9. Here the driving frequency $\omega = 0.95$ about equals the free oscillation frequency ω_0. The approach to steady state takes several oscillations, and while it is taking place the oscillation amplitude builds up to its steady-state value of approximately

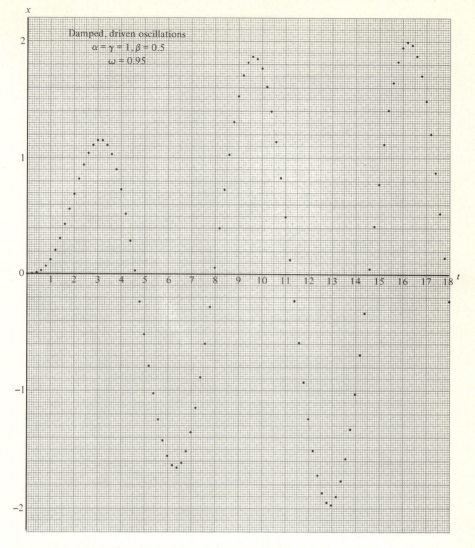

Figure 4-9. A damped oscillator driven at its resonant frequency.

2.0. At steady state the oscillation frequency again equals the driving frequency.

   Figure 4-10 shows the third example for which the driving frequency $\omega = 2.00$ is appreciably larger than the free-oscillation frequency $\omega_0 = 1$. In this case the approach to steady state takes even more oscillations, and the behavior of the oscillator seems quite chaotic for the first few oscillations.   ////

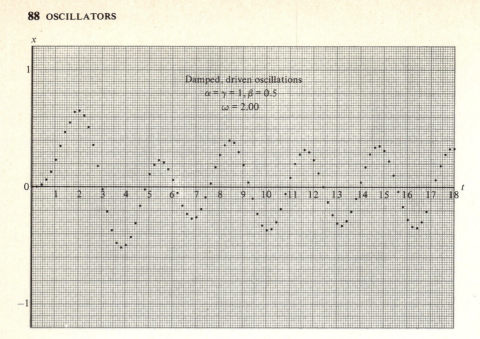

Figure 4-10. A damped oscillator driven at a frequency higher than its resonant frequency.

Why does the oscillator have such great difficulty in reaching steady state when ω is larger than ω_0, only moderate difficulty when ω approximately equals ω_0, and very little difficulty when ω is smaller than ω_0? [*Hint:* Is the effect (the motion of the oscillator) always in step (in phase) with the cause (the driving force acting on it)?]

4-5. RESONANCE

Steady-state amplitudes of an $\alpha = 1$, $\beta = 0.5$ oscillator and a $\gamma = 1$ driving force are plotted in Fig. 4-11 for the three values of driving frequency ω used in the example just considered. The figure also shows steady-state amplitudes obtained by running the program for a number of intermediate values of ω.

The points trace out the *resonance* curve for the $\alpha = 1$, $\beta = 0.5$ oscillator, and make it clear that the system exhibits a maximum response when the frequency of the applied force resonates with the frequency at which it would oscillate in the absence of such a force. Above or below resonance, the response to a driving force of the same amplitude is considerably smaller. This is certainly plausible if you think about what would happen if you were rhythmically pushing a child on a swing at a frequency higher than, equal to, or lower than the swing's free-oscillation frequency.

The resonance curve is not perfectly sharp because the oscillator is

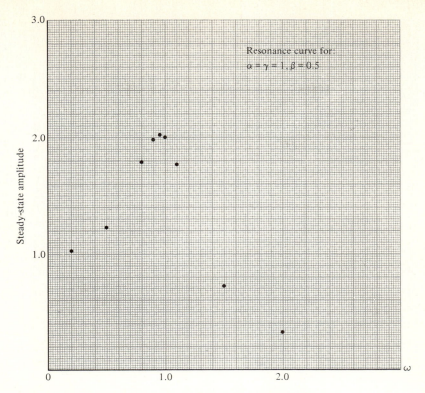

Figure 4-11. The resonance curve for a lightly damped oscillator.

damped. The effect is apparent if you look at results, plotted in Fig. 4-12, of runs made with $\beta = 0.25$. With half as much damping, the resonance peak is approximately twice as high and half as wide.

If you look carefully at both figures, you will note that the resonance peak occurs at a driving frequency somewhat smaller than the free-oscillation frequency $\omega_0 = 1$. The depression of the resonance frequency from the free-oscillation frequency is less pronounced the lighter the damping. Use physical considerations to explain why damping lowers the resonant frequency of the oscillator.

EXERCISES

4-1. State which electrical circuits correspond to the harmonic oscillator, the damped oscillator, and the damped, driven oscillator in that their behavior is governed by the same differential equations. How are the circuit parameters related to α and β?

4-2. Run the harmonic-oscillator program with the same α as in the ex-

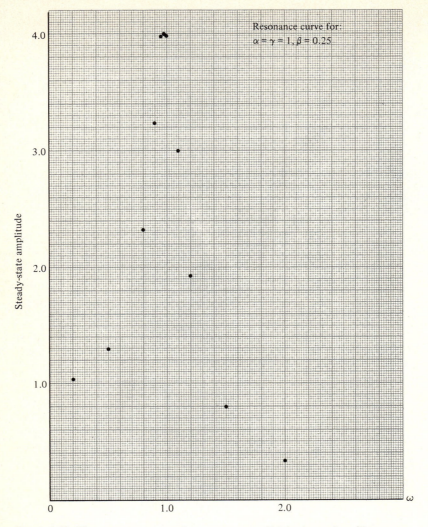

Figure 4-12. The resonance curve for an even more lightly damped oscillator.

ample, but with a variety of different initial values of x and dx/dt. Discuss your results.

4-3. Run the harmonic-oscillator program with several values of α to determine how the period T of the oscillation, or its reciprocal the frequency $\nu = 1/T$, depends on α. Plot your results for T or ν versus α, but do not bother to plot x versus t.

4-4. Run the large oscillation of a pendulum modification of the program, using the same α as in the example, with enough initial values of θ to determine how the period T or frequency ν depend on the amplitude of the oscillation. It is not fair for it to exceed $\pi/2$. Why? Plot your results for T or ν versus θ_{max}, but do not bother to plot θ versus t.

4-5. By running the damped-oscillator program with $\alpha = 1$ and various values of β, find the largest β for which x makes just one, imperceptibly small negative swing. Plot x versus t for this so-called critically damped case and compare it with the heavily damped case plotted in the example.

4-6. Use a value of β about 10 percent less than the value found in 4-5 and measure the period of the damped oscillations. How does it compare with the undamped oscillator having the same value $\alpha = 1$, and with the $\beta = 0.5$ damped oscillator with that value of α, that were treated in the examples?

4-7. Make the damped, driven oscillator program run the three sets of conditions shown in the example: $\alpha = \gamma = 1$, $\beta = 0.5$, $\omega = 0.95$; $\alpha = \gamma = 1$, $\beta = 0.5$, $\omega = 2.00$; $\alpha = \gamma = 1$, $\beta = 0.5$, $\omega = 0.20$. After steady state is achieved for each set, stop the program at a succession of values of t and record the values of $\sin \omega t$. Plot them over the plots of x shown in the example. Discuss the relations between the phase of the driving force and the phase of the response in each case. Can you see how this is related to the rapidity of the approach to steady state in each case?

4-8. Do the same for a driving force proportional to $\cos \omega t$ by simply changing the appropriate step in the program from sin to cos. Explain why the rapidity of approach to steady state is modified by doing this.

4-9. Make a set of damped, driven oscillator runs for $\alpha = \beta = \gamma = 1$, and various values of ω, to obtain a resonance curve for this case. It is not necessary to plot x versus t to determine the steady-state amplitude. Compare your results with those shown in the example, discussing particularly the height and width of the resonance curve and the location of its peak.

4-10. Do the same as in 4-9 for $\alpha = 1$, $\beta = 0.5$, $\gamma = 2$. Discuss your results in terms of the linearity of the differential equation.

4-11. Study the details of each program and explain precisely what happens in each step.

4-12. For the harmonic oscillator, all of the plots of x versus t look like sinusoidals. They are. Write a small program to generate a cosine curve of the appropriate amplitude and frequency, and plot some points on the plots of x shown in the example.

4-13. For oscillations of the pendulum, the plots of θ versus t also look like sinusoidals. But for large-amplitude oscillations they are not! Do the same as in 4-12. Consider the forces acting on the pendulum near the limits of its swing and interpret its behavior.

4-14. Modify the harmonic-oscillator program so that it will display for each loop the kinetic energy $K = m(dx/dt)^2/2$, the potential energy $V = kx^2/2$, and the total energy $E = K + V$ of the oscillator. Plot these for several oscillations and comment.

4-15. Write a program for an anharmonic oscillator (undamped and undriven) in which the restoring force is $- kx - px^3$. This describes an oscillator in which with increasing extension or compression the spring becomes more stiff if $p > 0$ and less stiff if $p < 0$. Run an example. There is no analytical solution for this system, or for the one in 4-16.

4-16. Do the same for the anharmonic restoring force $- kx + px^2$. For $p > 0$, this describes a spring which is stiffer when compressed a certain amount than when extended the same amount and can be used to model the oscillations in the center-to-center separation of the atoms of a diatomic molecule, as well as the way their average separation expands as the amplitude of the oscillation increases with increasing temperature. Make your program evaluate the average over a cycle of x and run several different amplitudes to study the thermal expansion of the molecule.

REFERENCES

1. Bueche, Frederick J.: *Introduction to Physics for Scientists and Engineers,* 2d ed., McGraw-Hill Book Company, New York, 1975, pp. 142 and 210.
    Halliday, David, and Robert Resnick: *Fundamentals of Physics,* John Wiley & Sons, Inc., New York, 1974, p. 101.
    Sears, Francis W., and Mark W. Zemansky: *University Physics,* 4th ed., Addison-Wesley Publishing Company, Inc., 1970, pp. 103 and 154.
    Weidner, Richard, and Robert Sells: *Elementary Classical Physics,* Allyn and Bacon, Inc., Boston, 1973, p. 142.
2. Halliday, David, and Robert Resnick: *Fundamentals of Physics,* John Wiley & Sons, Inc., New York, 1974, p. 628.
    Weidner, Richard, and Robert Sells: *Elementary Classical Physics,* Allyn and Bacon, Inc., Boston, 1973, p. 673.

COUPLED OSCILLATORS

5-1. INTRODUCTION

In this short chapter you will study the behavior of a system of two component oscillators which are coupled to each other by a spring connected between them. Perhaps you have seen a demonstration of such a system or one of its many equivalents in a high-school or college physics course. If an oscillation is set up in one of the components, with the other initially not moving, their coupling causes the oscillation to be transferred from the first to the second, then from the second to the first, and so on. But if both components of the system are initially set into oscillation in certain special ways, called normal modes, then the entire system will oscillate in very simple manners with each cycle repeating exactly the preceeding cycle.

Many mechanical, structural, and acoustical engineers spend much of their time analyzing systems of two or more coupled mechanical oscillators. The same is true for electrical engineers and systems of coupled electrical oscillators. Many physicists, particularly solid-state physicists, are concerned with systems consisting of a large number of coupled oscillators. And mathematicians are interested in the coupled differential equations that you will obtain and solve in the next section.

Another motivation for this chapter is that the extension of the half-increment method used to solve these coupled differential equations will be applied directly in the next chapter to solve the ones that arise in treating planetary motion and α-particle scattering.

5-2. COUPLED OSCILLATIONS

A system of two symmetrical *coupled oscillators* is shown in Fig. 5-1. It consists of two bodies of the same mass m, each connected to the ends of springs of the same stiffness constant k (see Eq. 4-1), the other ends of

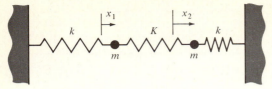

Figure 5-1. Coupled oscillators.

these springs being fixed. The two bodies are also connected to each other by a spring of stiffness constant K. When all three springs have their normal lengths, and the two bodies are in their corresponding equilibrium positions, the coordinates x_1 and x_2 specifying the locations of the two bodies are both equal to zero. Otherwise, the values of x_1 and x_2 give the locations of the bodies relative to their equilibrium locations. They also specify the extension or compression of the three springs. For instance, in the situation shown in the figure the spring connecting the body on the left to the support on the left is extended by the amount x_1, the spring connecting the two bodies is extended by the amount $(x_2 - x_1)$, and the spring connecting the body on the right to the support on the right is compressed by the amount x_2.

There will be two differential equations for this system, one for x_1 and the other for x_2. The first is obtained by applying Newton's law to the body on the left

$$\frac{d^2x_1}{dt^2} = \frac{F_1}{m} = \frac{-kx_1 + K(x_2 - x_1)}{m} \tag{5-1}$$

The quantity F_1 is the total force acting on the body on the left whose mass is m and acceleration is d^2x_1/dt^2. Again taking the situation illustrated in Fig. 5-1, you can see that one contribution to F_1 is the force $-kx_1$ caused by the spring connecting it to the support on the left. Its magnitude is kx_1 because the extension of the spring is x_1, and it is negative since that spring applies its force in the direction of negative x_1. The other contribution to F_1, namely $+K(x_2 - x_1)$, arises from the force applied to the body on the left by the interconnecting spring, which is extended by the amount $(x_2 - x_1)$ and so is pulling the body in the direction of positive x_1 with a force of strength $K(x_2 - x_1)$.

You should go through a similar analysis of the forces acting on the body on the right and thereby convince yourself that the differential equation for the coordinate x_2 is

$$\frac{d^2x_2}{dt^2} = \frac{F_2}{m} = \frac{-kx_2 - K(x_2 - x_1)}{m} \tag{5-2}$$

Introducing the parameters

$$\alpha = \frac{k + K}{m} \qquad \beta = \frac{K}{m} \tag{5-3}$$

you can immediately convert (5-1) and (5-2) into the forms

$$\frac{d^2x_1}{dt^2} = -(\alpha x_1 - \beta x_2) = C_1 \qquad (5\text{-}4)$$

and

$$\frac{d^2x_2}{dt^2} = -(\alpha x_2 - \beta x_1) = C_2 \qquad (5\text{-}5)$$

These forms allow you to write, by inspection of Eqs. (3-17), (3-18), and (3-19), the basic equations you need for their half-increment method solution. There will be six such equations, three in x_1 and three in x_2,

$$\frac{dx_1}{dt_{1/2}} \simeq \frac{dx_1}{dt_0} + C_1\frac{\Delta t}{2} \qquad\qquad C_1 = -(\alpha x_{1_0} - \beta x_{2_0}) \qquad (5\text{-}6)$$

$$\frac{dx_1}{dt_{i+1/2}} \simeq \frac{dx_1}{dt_{i-1/2}} + C_1\Delta t \qquad\qquad C_1 = -(\alpha x_{1_i} - \beta x_{2_i}) \qquad (5\text{-}7)$$

$$x_{1_{i+1}} \simeq x_{1_i} + \frac{dx_1}{dt_{i+1/2}}\Delta t \qquad (5\text{-}8)$$

and

$$\frac{dx_2}{dt_{1/2}} \simeq \frac{dx_2}{dt_0} + C_2\frac{\Delta t}{2} \qquad\qquad C_2 = -(\alpha x_{2_0} - \beta x_{1_0}) \qquad (5\text{-}9)$$

$$\frac{dx_2}{dt_{i+1/2}} \simeq \frac{dx_2}{dt_{i-1/2}} + C_2\Delta t \qquad\qquad C_2 = -(\alpha x_{2_i} - \beta x_{1_i}) \qquad (5\text{-}10)$$

$$x_{2_{i+1}} \simeq x_{2_i} + \frac{dx_2}{dt_{i+1/2}}\Delta t \qquad (5\text{-}11)$$

The scheme for using these equations to obtain a numerical solution to the coupled differential equations is shown in Fig. 5-2. At first glance it may seem terribly complicated, but if you spend a minute with it you will see that it is nothing more than a superposition of two schemes of the general form shown in Fig. 3-7, one for x_1 and the other for x_2, plus interconnecting arrows that represent the coupling. For instance, the arrow leading from x_{1_0} to $dx_2/dt_{1/2}$ shows that the value of the latter depends on the value of the former because C_2 in Eq. (5-9) contains x_{1_0}.

You will find a program for carrying out the scheme on your calculator in Tables 5-1 (HP-25) or (SR-56).

For the HP-25
Switch to **PRGM**, clear by pressing **f PRGM**, then key steps 01 through 33. Step 34 can be either a pause or a stop instruction; you might try **f PAUSE** at first. After next keying steps 35 through 39, you again come to a choice of display and running modes; try **g NOP** at first. Then key the remainder of the program, switch to **RUN**, and key **f PRGM** to set to step 00. Preload registers 0 through 6 with the four initial values of position and speed you want to use and with the two values of the parameters specifying the spring-stiffness-to-mass ratios, in the usual way. Push **R/S** to start and you will be watching x_1, the coordinate of the body on the left. Soon you will

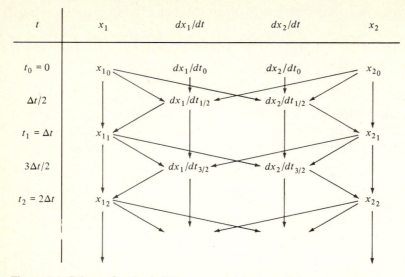

t	x_1	dx_1/dt	dx_2/dt	x_2

Figure 5-2. Scheme for the half-increment method applied to the solution of coupled differential equations. If the C coefficients contain functions of t (not so for systems treated in this book), then each t_i will feed into $dx_1/dt_{i+1/2}$ and $dx_2/dt_{i+1/2}$.

also want to look at x_2. To do so, push **R/S** to stop, key **GTO 3 9,** switch to **PRGM,** key **f PAUSE,** switch to **RUN,** key **f PRGM,** reload registers 0 through 3, then push **R/S** to run. For plotting you may either have both steps 34 and 40 be **R/S,** or have 34 be **f PAUSE** and 40 be **R/S.** The second way is more efficient, but takes some concentration. To see the value of t any time the calculator is stopped by a program **R/S,** or by an **R/S** executed manually during a program **f PAUSE,** key **RCL 7.**

For the SR-56
Clear by keying **2nd CP,** key **LRN,** then key steps 00 through 64. Steps 65 and 66 can either be **2nd pause 2nd pause** or be **R/S 2nd pause;** you might try the pause instruction first. After next keying steps 67 through 76, you again come to a choice of display and running modes; try **2nd NOP 2nd NOP** at first. Then key the remainder of the program, key **LRN,** and key **RST** to set to step 00. Preload registers 0 through 6 with the four initial values of position and speed you want to use and with the two values of the parameters specifying the spring-stiffness-to-mass ratios, in the usual way. Push **R/S** to start and you will be watching x_1, the coordinate of the body on the left. Soon you will also want to look at x_2. To do so, push **R/S** to stop, key **GTO 7 7,** key **LRN,** key **2nd pause,** key **2nd pause,** key **LRN,** key **RST,** reload registers 0 through 3, then push **R/S** to run. For plotting you may either have both steps 65–66 and 77–78 be **R/S 2nd NOP,** or have the first pair be **2nd pause 2nd pause** and the second be **R/S 2nd**

NOP. The latter is more efficient, but takes some concentration. To see the value of t any time the calculator is stopped by a program **R/S**, key **RCL 7**. If it is stopped by a **R/S** executed manually after a pause, key **2nd EXC 7**, inspect t, then again key **2nd EXC 7**.

Example

initial $x_1 = 1$, initial $dx_1/dt = 0$, initial $x_2 = 0$,
initial $dx_2/dt = 0$, $\Delta t = 0.4$, $\alpha = 1.25$, $\beta = 0.25$

The values used for α and β correspond to $k/m = 1$ and $K/m = 0.25$; that is, the spring that does the coupling is less stiff by a factor of $\frac{1}{4}$ than the other two springs. The system was started with the body on the left displaced by a unit distance to the right and released without giving it any initial speed, while the body on the right was given no initial displacement or speed. What happened next can be seen in Fig. 5-3.

It is interesting to observe how the oscillation in x_1 dies out while the oscillation in x_2 builds up until, at $t \simeq 13$, the motion in the system is completely transferred to x_2. Then as time continues to unfold, the process reverses itself until the oscillation has flowed back to x_1, and so on forever (in the absence of friction).   ////

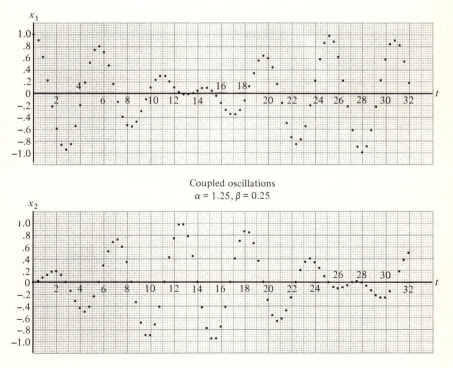

Figure 5-3. Typical behavior of a system of coupled oscillators.

Table 5-1. (HP-25) Coupled oscillations

Register Contents:

0	1	2	3	4	5	6	7
x_1	dx_1/dt	x_2	dx_2/dt	Δt	α	β	t
preloaded							

Program

Step	Code	Key Entry	X	Y	Z	T	Comments
00	00	0	0				7 zeroed
01	23 07	STO 7	0				Start loop
02	24 05	RCL 5	α				
03	24 02	RCL 2	x_2	α			
04	61	×	αx_2				
05	24 06	RCL 6	β	αx_2			
06	24 00	RCL 0	x_1	β	αx_2		
07	61	×	βx_1	αx_2			
08	41	−	$-C_2$				$C_2 = -(\alpha x_2 - \beta x_1)$
09	24 05	RCL 5	α	$-C_2$			
10	24 00	RCL 0	x_1	α	$-C_2$		
11	61	×	αx_1	$-C_2$			
12	24 06	RCL 6	β	αx_1	$-C_2$		
13	24 02	RCL 2	x_2	β	αx_1	$-C_2$	
14	61	×	βx_2	αx_1	$-C_2$		
15	41	−	$-C_1$	$-C_2$			$C_1 = -(\alpha x_1 - \beta x_2)$
16	24 07	RCL 7	t	$-C_1$	$-C_2$		
17	15 71	g x = 0	t	$-C_1$	$-C_2$		Test t for routing to put 1/2 in preliminary loop
18	13 42	GTO 42	t	$-C_1$	$-C_2$		
19	22	R↓	$-C_1$	$-C_2$	$-C_2$		
20	24 04	RCL 4	Δt	$-C_1$	$-C_2$		Will be $-C_1/2$, $-C_2/2$ in prelim. loop
21	61	×	$-C_1 \Delta t$	$-C_2$			1 now $dx_1/dt_{1/2}$ in prelim. loop
22	23 41 01	STO − 1	$-C_1 \Delta t$	$-C_2$			1 now $dx_1/dt_{3/2}$ in next loop

No.	Code	Keystroke	X	Y	Z	Comment
24	22	R↓ *(Z)*	$-C_2$			
25	24 04	RCL 4	Δt	$-C_2$		
26	61	x	$-C_2\Delta t$			3 now $dx_2/dt_{1/2}$ in prelim. loop
27	23 41 03	STO – 3	$-C_2\Delta t$			3 now $dx_2\,dt_{3/2}$ in next loop
28	24 04	RCL 4	Δt	Δt		7 now t_1 in prelim. loop
29	23 51 07	STO + 7	Δt			7 now t_2 in next loop
30	24 01	RCL 1	dx_1/dt			
31	61	x	$\Delta t\,dx_1/dt$			0 now x_{1_1} in prelim. loop
32	23 51 00	STO + 0	$\Delta t\,dx_1/dt$			0 now x_{1_2} in next loop
33	24 00	RCL 0	x_1			
34	14 74 (or 74)	f PAUSE (or R/S)	x_1			
35	24 04	RCL 4	Δt	Δt		
36	24 03	RCL 3	dx_2/dt			
37	61	x	$\Delta t\,dx_2/dt$			2 now x_{2_1} in prelim. loop
38	23 51 02	STO + 2	$\Delta t\,dx_2/dt$			2 now x_{2_2} in next loop
39	24 02	RCL 2	x_2			
40	14 74 (or 15) (or 74)	f PAUSE (or g NOP) (or R/S)	x_2			
41	13 03	GTO 03 *(0 3)*	x_2			To new loop
42	22	R↓	$-C_1$	$-C_2$		
43	02	2	2	$-C_1$		
44	71	÷	$-C_1/2$	$-C_2$		
45	21	x ≷ y	$-C_2$	$-C_1/2$	$-C_2$	
46	02	2	2	$-C_2$		
47	71	÷	$-C_2/2$	$-C_1/2$		
48	21	x ≷ y	$-C_1/2$	$-C_2/2$		
49	13 21	GTO 21 *(SUB 7 1)*	$-C_1/2$	$-C_2/2$	$-C_1/2$	To continue loop

Table 5-1. (SR-56) Coupled oscillations

Register Contents:

0	1	2	3	4	5	6	7	8	9
x_1	dx_1/dt	x_2	dx_2/dt	Δt	α	β	t	$C_1, C_1/2$	$C_2, C_2/2$
			preloaded						

Program

Step	Code	Key Entry	Comments
00	00	0	0
01	33	STO	
02	07	7	7 zeroed
03	56	2nd CP	Test register zeroed
04	34	RCL	Start loop 5∪B2
05	06	6	β
06	64	×	
07	34	RCL	
08	02	2	x_2
09	74	−	
10	34	RCL	SUB 1
11	05	5	α
12	64	×	
13	34	RCL	
14	00	0	x_1
15	94	=	$C_1 = \beta x_2 - \alpha x_1$
16	33	STO	SUB 1 (+ 0 =) }
17	08	8	8 now C_1
18	34	RCL	β SUB2
19	06	6	
20	64	×	
21	34	RCL	
22	00	0	x_1
23	74	−	
24	34	RCL	

Step	Code	Key Entry	Comments
51	35	SUM	3 now $dx_2/dt_{1/2}$ in prelim. loop, $dx_2/dt_{3/2}$ in next
52	03	3	
53	34	RCL	Δt
54	04	4	
55	35	SUM	7 now t_1 in prelim. loop, t_2 in next
56	07	7	
57	64	×	$dx_1/dt_{1/2}$ in prelim. loop, $dx_1/dt_{3/2}$ in next
58	34	RCL	
59	01	1	$\Delta t\, dx_1/dt_{1/2}$ in prelim. loop, $\Delta t\, dx_1/dt_{3/2}$ in next
60	94	=	
61	35	SUM	0 now x_{1_1} in prelim. loop, x_{1_2} in next
62	00	0	
63	34	RCL	x_{1_1} in prelim. loop, x_{1_2} in next
64	00	0	
65	59 (or 41)	2nd PAUSE (or R/S)	
66	59 (or 46)	2nd PAUSE (or 2nd NOP)	
67	34	RCL	Δt
68	04	4	

Step	Code	Key	Comment
25	05	5	
26	64	×	
27	34	RCL	α *SUB 1*
28	02	2	
29	94	=	$x_2 = \beta x_1 - \alpha x_2$
30	33	STO	*SUB 2 (+ 0 =)*
31	09	9	9 now C_2
32	34	RCL	
33	07	7	t
34	37	2nd x = t	Test t for routing in preliminary loop
35	08	8	
36	02	2	
37	34	RCL	Δt
38	04	4	
39	64	×	
40	34	RCL	*S1*
41	08	8	$C_1/2$ in prelim. loop, C_1 in next loop
42	94	=	$\Delta t C_1/2$ in prelim. loop, $\Delta t C_1$ in next loop
43	35	SUM	1 now $dx_1/dt_{1/2}$ in prelim. loop, $dx_1/dt_{3/2}$ in next loop
44	01	1	
45	34	RCL	Δt
46	04	4	
47	64	×	
48	34	RCL	*S2*
49	09	9	$C_2/2$ in prelim. loop, C_2 in next loop
50	94	=	$\Delta t C_2/2$ in prelim. loop, $\Delta t C_2$ in next loop

Step	Code	Key	Comment
69	64	×	
70	34	RCL	
71	03	3	$dx_2/dt_{1/2}$ in prelim. loop, $dx_2/dt_{3/2}$ in next
72	94	=	$\Delta t\, dx_2/dt_{1/2}$ in prelim. loop, $\Delta t\, dx_2/dt_{3/2}$ in next
73	35	SUM	
74	02	2	2 now x_{2_1} in prelim. loop, x_{2_2} in next
75	34	RCL	
76	02	2	x_{2_1} in prelim. loop, x_{2_2} in next
77	59 (or 46) (or 41)	2nd PAUSE (or 2nd NOP) (or R/S)	
78	59 (or 46)	2nd PAUSE (or 2nd NOP)	
79	22	GTO	*0*
80	00	0	
81	04	4	To new loop
82	02	2	2
83	12	INV	
84	30	2nd PROD	
85	08	8	8 now $C_1/2$
86	12	INV	
87	30	2nd PROD	
88	09	9	9 now $C_2/2$
89	22	GTO	*3*
90	03	3	
91	07	7	To continue loop

It is possible to construct a physical explanation of the general behavior of coupled oscillators by superposing two discussions similar to that given in Sec. 4-2 for a single oscillator, providing account is taken of the interaction caused by the presence of the coupling spring. The same can be done to obtain a mathematical explanation of the general behavior of the solutions to the coupled-oscillator differential equations. But these explanations are too complicated to be of much use.

So take the experimental approach. Try running the program with different values of K, noting how K affects the rapidity of transfer of the motion from x_1 to x_2, and back. Also try running with different initial conditions, and perhaps you will discover ones that produce a particularly simple behavior in which there is no flow of motion from one body to the other because they both oscillate in each cycle in exactly the same way they oscillated in the previous one. If you find such behavior, you have found a normal mode of the system.

5-3. NORMAL MODES

Two special sets of initial conditions are used in the following examples.

Examples

initial $x_1 = 1$, initial $dx_1/dt = 0$, initial $x_2 = 1$,
initial $dx_2/dt = 0$, $\Delta t = 0.4$, $\alpha = 1.25$, $\beta = 0.25$

initial $x_1 = 1$, initial $dx_1/dt = 0$, initial $x_2 = -1$,
initial $dx_2/dt = 0$, $\Delta t = 0.4$, $\alpha = 1.25$, $\beta = 0.25$

The behavior of x_1 and x_2 in the first example can be seen in Fig. 5-4. Figure 5-5 plots these coordinates versus the time t for the second example. Do you see what is meant by saying that the system has a particularly simple behavior when it is in a *normal mode*?

Note that the period of the oscillations in x_1 and x_2 in the normal mode shown in Fig. 5-4 is longer than the period of the oscillations in the normal mode shown in Fig. 5-5. If you think of what the interconnecting spring is doing in each of these, you will understand why the period is longer, or the frequency is lower, in the first normal mode. ////

Can you find a third normal mode for the system? Before giving a superficial answer, answer the next question: Are the normal modes obtained by using the initial conditions $x_1 = -1$, $dx_1/dt = 0$, $x_2 = -1$, $dx_2/dt = 0$, or the initial conditions $x_1 = -1$, $dx_1/dt = 0$, $x_2 = 1$, $dx_2/dt = 0$, really different from the two modes found in the examples, for this system governed by linear differential equations?

The physical properties of the system when oscillating in the two normal modes illustrated in Figs. 5-4 and 5-5 can be exploited to simplify the

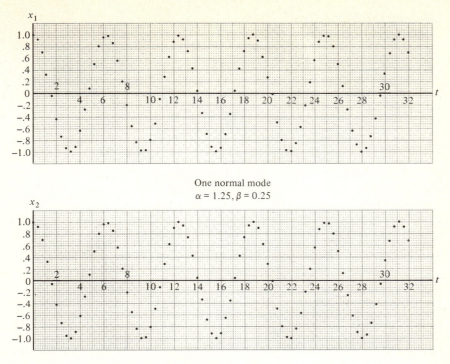

One normal mode
$\alpha = 1.25, \beta = 0.25$

Figure 5-4. Coupled oscillators in one normal mode.

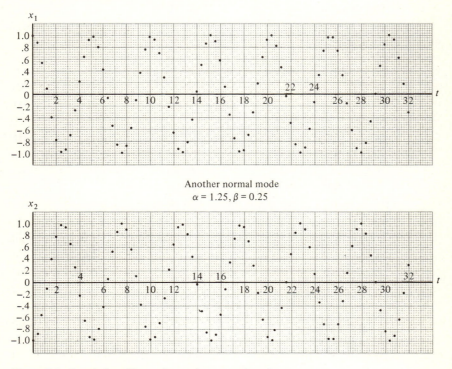

Another normal mode
$\alpha = 1.25, \beta = 0.25$

Figure 5-5. Coupled oscillators in another normal mode.

mathematical treatment of the coupled oscillator by using as coordinates not x_1 and x_2 but, instead, two different combinations of these quantities. Use the physical intuition that you have gained in this chapter to suggest appropriate forms for the two new coordinates and, if you are mathematically inclined, reformulate the two differential equations in terms of them.

EXERCISES

5-1. State which electrical circuit corresponds to the coupled oscillator. How are the circuit parameters related to α and β?

5-2. Run the coupled-oscillator program with the same parameters $\alpha = 1.25$, $\beta = 0.25$, or $k/m = 1$, $K/m = 0.25$, as are used in the example, but with different initial values of x_1, dx_1/dt, x_2, dx_2/dt. For instance, try $x_2 = 1$, $x_1 = dx_1/dt = dx_2/dt = 0$, and also try $x_1 \neq 0$, $x_2 \neq 0$, with $x_1 \neq x_2$, $dx_1/dt = dx_2/dt = 0$. Observe and describe the oscillations without bothering to plot them.

5-3. See if it is possible to find a set of initial conditions which will produce a third normal mode for the system, i.e., an oscillation in which for both x_1 and x_2 each oscillation exactly repeats the preceding oscillation, but which is different from the two normal modes shown in the example.

5-4. Run the program to study the behavior of the coupled oscillator with various initial conditions using several different sets of α and β, or k/m and K/m. In particular, investigate making K/k large and then making K/k small. For each value of K/k, measure the frequencies of both normal modes by stopping the calculator at the end of each successive cycle of x_1 and of x_2 and recording the values of t. Obtain enough data to make a plot of the frequencies of both normal modes versus K/k for a fixed m, say $m = 1$. Explain your results.

5-5. Repeat 5-4 for $K/k = 0$ and then for $K/k = \infty$ (that is, set $k = 0$). Describe and explain the behavior of the system in both these limits, stating just what physical situation each limit pertains to. What is the relation of the $K/k = \infty$ limit to the motion of a diatomic molecule?

5-6. Study the details of the program and explain precisely what happens in each step.

5-7. The normal modes look like sinusoidals, and they are. Write a small program to generate two sinusoidal curves of appropriate amplitudes and frequencies, and plot some points on the plots of x_1 and x_2 for the two normal modes shown in the example.

5-8. Using the procedure of 5-7, make the program generate one-half the sum of the values of x_1, or of x_2, for the two sinusoidal curves you fitted to the normal modes. Plot some points on plots shown in the example for the oscillation with initial values $x_1 = 1$, $x_2 = dx_1/dt = dx_2/dt = 0$. Now explain the relation between the normal modes and the typical oscillation studied in the example.

5-9. Using the same $\alpha = 1.25$ and $\beta = 0.25$ used in the example, plot x_1 and x_2 for the initial conditions $x_2 = 1$, $x_1 = dx_1/dt = dx_2/dt = 0$. What is the relation of this oscillation to the normal modes? Modify the procedure of 5-8 to demonstrate the relation.

SIX

CENTRAL FORCE MOTION

6-1. INTRODUCTION

It would be reasonable to say that the most important problem in physics is that of planetary motion. Many historians consider the field of physics to date from the work of Newton, and the motion of the planets was the principal problem Newton set out to solve. As you may know, in the process of doing this he not only introduced his law of motion and discovered the law of gravity, he also independently developed differential and integral calculus. So planetary motion has had an equally important place in the history of mathematics.

In working through this chapter, you too will solve the problem of planetary motion, including the case of unbound trajectories that has such contemporary significance to NASA. Furthermore, you will solve the closely related α-particle scattering problem which played the key role in the discovery of the nucleus of the atom.

The method used for solving the differential equations that arise will be taken directly from the extension of the general half-increment method that was developed in the preceding chapter. It is interesting to point out that the mathematical procedures employed in the method are more closely related, at least in spirit, to those Newton himself employed than are the procedures of the modern analytical methods. Once he got used to the calculator, Newton would have felt very much at home here.

6-2. CENTRAL FORCE EQUATIONS AND SOLUTIONS

Central force motion is the motion of a body under the influence of an attractive (or repulsive) force acting always to (or from) a point located at the center of the coordinate system used to describe its motion (see Fig. 6-1). Examples involving attractive forces are the motion of a planet about the sun or a satellite about the earth (the force is gravitational), and an electron about a proton in a hydrogen atom (the force is electrical). These are examples of central force motion if two approximations are made. The first is that you can ignore the forces acting between the mov-

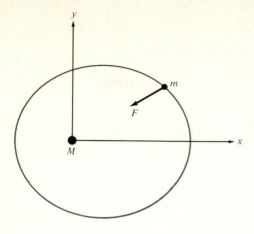

Figure 6-1. Illustration of central-force motion for the case of an attractive force of magnitude F acting on a moving body (e.g., the planet Mercury), of mass m. The direction of the force is always toward the center of the coordinate system where the other body (e.g., the sun), of *very* much larger mass M, can be considered to be fixed.

ing body and everything except the body fixed at the center of coordinates. The second is that this body can actually be considered as fixed. Both approximations are very well justified for planetary motion since the sun is so massive. This means that the gravitational force acting between it and any planet dominates the gravitational forces acting between the planets, and also that the sun can be taken to be fixed at the center of an appropriately chosen coordinate system. The approximations are also well justified for satellite motion.

They are a little less justified for the motion of an electron in a hydrogen atom, since the proton mass M is only about 2000 times larger than the electron mass m, and so the proton cannot be precisely stationary. But it is easy to correct for this because it turns out that the body of mass m moves relative to the body of mass M as though the latter were fixed and the mass of the former were reduced slightly to the value $mM/(m + M)$. (See Ref. 1.)

An example of central force motion involving a repulsive force (providing a reduced mass correction is made) is the motion of a positively charged α particle when it scatters from a positively charged nucleus due to the electrical force acting between them.

Now, consider the equations. This two-dimensional problem will involve two of them, one for the x coordinate of the moving body and the other for its y coordinate. Applying Newton's law of motion separately to these coordinates gives

$$\frac{d^2x}{dt^2} = \frac{F_x}{m} \tag{6-1}$$

and

$$\frac{d^2y}{dt^2} = \frac{F_y}{m} \tag{6-2}$$

Here F_x and F_y are the x and y components of the force acting on the

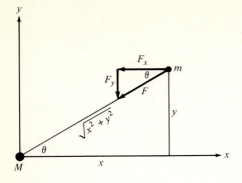

Figure 6-2. Resolution of the force acting on a moving body into its x and y components for the case of an attractive force.

body. Figure 6-2 shows the relations between these components, the magnitude F of the total force, and the angle θ between the x axis and a line drawn from the origin to the body, for the case of an attractive force. You can see from the figure that

$$F_x = -F\cos\theta \qquad F_y = -F\sin\theta$$

and also that

$$\cos\theta = \frac{x}{\sqrt{x^2 + y^2}} \qquad \sin\theta = \frac{y}{\sqrt{x^2 + y^2}}$$

Therefore

$$F_x = -F\frac{x}{\sqrt{x^2 + y^2}} \qquad F_y = -F\frac{y}{\sqrt{x^2 + y^2}}$$

Using these in Eqs. (6-1) and (6-2) you obtain

$$\frac{d^2x}{dt^2} = -\frac{F}{m}\frac{x}{\sqrt{x^2 + y^2}} \tag{6-3}$$

and

$$\frac{d^2y}{dt^2} = -\frac{F}{m}\frac{y}{\sqrt{x^2 + y^2}} \tag{6-4}$$

For a gravitational force, Newton's law of gravitation (Ref. 2) says that the magnitude of F is

$$F = GmM\frac{1}{(x^2 + y^2)} \tag{6-5}$$

where G is the gravitational constant. For an electrical force, Coulomb's law of electrostatics (Ref. 3) says that its magnitude is

$$F = \frac{qQ}{4\pi\epsilon_0}\frac{1}{(x^2 + y^2)} \tag{6-6}$$

where q and Q are the magnitudes of the charges of the moving and fixed bodies, and where $1/4\pi\epsilon_0$ is the Coulomb's law constant. If the two charges are of opposite sign, as for an electron and a nucleus, then the force is attractive and Eq. (6-6) can be applied directly to (6-3) and (6-4). But if they are of the same sign, as for an α particle and a nucleus, then when applying Eq. (6-6), the minus signs in (6-3) and (6-4) must be changed to plus signs to account for the fact that the force will be repulsive.

Putting it all together, you can write the two differential equations as

$$\frac{d^2x}{dt^2} = -\frac{\alpha x}{(x^2 + y^2)^{3/2}} = -\alpha\,(x^2 + y^2)^{\beta}\,x = C_x \tag{6-7}$$

and

$$\frac{d^2y}{dt^2} = -\frac{\alpha y}{(x^2 + y^2)^{3/2}} = -\alpha\,(x^2 + y^2)^{\beta}\,y = C_y \tag{6-8}$$

where $\qquad \alpha = \begin{cases} GM, \text{ for a gravitational force} \\ qQ/4\pi\epsilon_0 m, \text{ for an attractive electrical force} \\ -qQ/4\pi\epsilon_0 m, \text{ for a repulsive electrical force} \end{cases} \tag{6-9}$

and where $\beta = -3/2$, for a gravitational or electrical force $\qquad$ (6-10)

The reason for writing the equations in terms of β, instead of simply making the $-\frac{3}{2}$ power explicit, is that the program will allow for the possibility of varying β and thus studying some interesting things that would happen if the law of gravitation, for instance, was not an inverse-square law.

If you look at Eqs. (6-7) and (6-8) from the perspective of the previous chapter, you will immediately recognize that the problem of central force motion has led you again to a pair of coupled differential equations. You should be able to solve them in the same way you did in the last chapter.

Indeed, you can! All that is required is to rewrite Eqs. (5-6) through (5-11) by changing x_1 to x, x_2 to y, C_1 to $C_x = -\alpha\,(x^2 + y^2)^{\beta}\,x$, and C_2 to $C_y = -\alpha\,(x^2 + y^2)^{\beta}\,y$. The scheme to be used is identical to the one in Fig. 5-2, except for changing x_1 to x and x_2 to y.

It is almost as easy to rewrite the programs of the last chapter. If you look at the results shown in Tables 6-1 (SR-56) or (HP-25), you will see that it is just a matter of changing from x_1, x_2 to x, y, and also changing the routine for generating the C coefficients. The programs are entered and used in exactly the same way as those of Chap. 5, except that there is never any point in running without displaying both x and y.

6-3. ORBITS

You can put your calculator in orbit by operating the program with one of the sets of conditions shown in the following examples.

Examples

initial $x = 1$, initial $dx/dt = 0$, initial $y = 0$,
initial $dy/dt = 1.0$, $\Delta t = 0.1$, $\alpha = 1$, $\beta = -1.5$

initial $x = 1$, initial $dx/dt = 0$, initial $y = 0$,
initial $dy/dt = 1.1$, $\Delta t = 0.1$, $\alpha = 1$, $\beta = -1.5$

Table 6-1. (SR-56) Central force motion

Register Contents:

0	1	2	3	4	5	6	7	8	9
x	dx/dt	y	dy/dt	Δt	α	β	t	$C_x, C_x/2$	$C_y, C_y/2$
		preloaded							

Program

Step	Code	Key Entry	Comments
00	00	0	0
01	33	STO	
02	07	7	7 zeroed
03	56	2nd CP	Test register zeroed
04	34	RCL	Start loop
05	00	0	x
06	43	x^2	x^2
07	84	+	
08	34	RCL	
09	02	2	y
10	43	x^2	y^2
11	94	=	$x^2 + y^2$
12	45	y^x	
13	34	RCL	
14	06	6	β
15	94	=	$(x^2 + y^2)^\beta$
16	64	×	
17	34	RCL	
18	05	5	α
19	94	=	$\alpha(x^2 + y^2)^\beta$
20	93	+/-	$-\,\alpha(x^2 + y^2)^\beta$
21	33	STO	
22	08	8	8 now $-\,\alpha(x^2 + y^2)^\beta$

Step	Code	Key Entry	Comments
51	94	=	$\Delta t C_y/2$ in prelim. loop, $\Delta t C_y$ in next loop
52	35	SUM	
53	03	3	3 now $dy/dt_{1/2}$ in prelim. loop, $dy/dt_{3/2}$ in next
54	34	RCL	
55	04	4	Δt
56	35	SUM	
57	07	7	7 now t_1 in prelim. loop, t_2 in next
58	64	×	
59	34	RCL	
60	01	1	$dx/dt_{1/2}$ in prelim. loop, $dx/dt_{3/2}$ in next
61	94	=	$\Delta t\, dx/dt_{1/2}$ in prelim. loop, $\Delta t\, dx/dt_{3/2}$ in next
62	35	SUM	
63	00	0	0 now x_1 in prelim. loop, x_2 in next
64	34	RCL	
65	00	0	
66	59 (or 41)	2nd PAUSE (or R/S)	x_1 in prelim. loop. x_2 in next

Step	Code	Instruction	Comment
23	33	STO	
24	09	9	9 now $-\alpha(x^2 + y^2)^\beta$
25	34	RCL	
26	00	0	x
27	30	2nd PROD	
28	08	8	8 now C_x
29	34	RCL	
30	02	2	y
31	30	2nd PROD	
32	09	9	9 now C_y
33	34	RCL	
34	07	7	t
35	37	2nd x = t	Test t for routing in preliminary loop
36	08	8	
37	03	3	
38	34	RCL	
39	04	4	Δt
40	64	×	
41	34	RCL	
42	08	8	$C_x/2$ in prelim. loop, C_x in next loop
43	94	=	$\Delta t C_x/2$ in prelim. loop, $\Delta t C_x$ in next loop
44	35	SUM	
45	01	1	1 now $dx/dt_{1/2}$ in prelim. loop, $dx/dt_{3/2}$ in next
46	34	RCL	
47	04	4	Δt
48	64	×	
49	34	RCL	
50	09	9	$C_y/2$ in prelim. loop, C_y in next loop

Step	Code	Instruction	Comment
67	59 (or 46)	2nd PAUSE (or 2nd NOP)	
68	34	RCL	
69	04	4	Δt
70	64	×	
71	34	RCL	
72	03	3	$dy/dt_{1/2}$ in prelim. loop, $dy/dt_{3/2}$ in next
73	94	=	$\Delta t \, dy/dt_{1/2}$ in prelim. loop, $\Delta t \, dy/dt_{3/2}$ in next
74	35	SUM	
75	02	2	2 now y_1 in prelim. loop, y_2 in next
76	34	RCL	
77	02	2	y_1 in prelim. loop, y_2 in next
78	59 (or 41)	2nd PAUSE (or R/S)	
79	59 (or 46)	2nd PAUSE (or 2nd NOP)	
80	22	GTO	
81	00	0	To new loop
82	04	4	
83	02	2	2
84	12	INV	
85	30	2nd PROD	
86	08	8	8 now $C_x/2$
87	12	INV	
88	30	2nd PROD	
89	09	9	9 now $C_y/2$
90	22	GTO	
91	03	3	
92	08	8	To continue loop

Table 6-1. (HP-25) Central force motion

Register Contents:

0	1	2	3	4	5	6	7
x	dx/dt	y	dy/dt	Δt	α	β	t

preloaded

Program

Step	Code	Key Entry	X	Y	Z	T	Comments
00							
01	00	0	0				
02	23 07	STO 7	0				7 zeroed
03	24 02	RCL 2	y				Start loop
04	24 02	RCL 2	y	y			
05	15 02	g x^2	y^2	y			
06	24 00	RCL 0	x	y^2	y		
07	15 02	g x^2	x^2	y^2	y		
08	51	+	y^2+x^2	y			
09	24 06	RCL 6	β	y^2+x^2	y		
10	14 03	f y^x	$(y^2+x^2)^\beta$	y			
11	24 05	RCL 5	α	$(y^2+x^2)^\beta$	y		
12	61	$\times$	$(y^2+x^2)^\beta\alpha$	y			$C_y = -\alpha(x^2+y^2)^\beta\, y$
13	61	$\times$	$-C_y$				
14	14 73	f LAST x	$(y^2+x^2)^\beta\alpha$	$-C_y$			
15	24 00	RCL 0	x	$(y^2+x^2)^\beta\alpha$	$-C_y$		
16	61	$\times$	$-C_x$	$-C_y$			$C_x = -\alpha(x^2+y^2)^\beta\, x$
17	24 07	RCL 7	t	$-C_x$	$-C_y$		
18	15 71	g $x=0$	t	$-C_x$	$-C_y$		Test t for routing to put 1/2 in preliminary loop
19	13 42	GTO 42	t	$-C_x$	$-C_y$		
20	22	R↓	$-C_x$	$-C_y$			
21	24 04	RCL 4	Δt	$-C_x$	$-C_y$		Will be $-C_x/2$, $-C_y/2$ in prelim. loop
22	61	$\times$	$-C_x\Delta t$	$-C_y$			

Step	Code			Instruction				Comment
23	23	41	01	STO − 1	$-C_x \Delta t$	$-C_y$		1 now $dx/dt_{1/2}$ in prelim. loop / 1 now $dx/dt_{3/2}$ in next loop
24	22			R ↓	$-C_y$	Δt		
25	24	04		RCL 4	Δt			
26	61			x	$-C_y \Delta t$			
27	23	41	03	STO − 3	$-C_y \Delta t$			
28	24	04		RCL 4	Δt			3 now $dy/dt_{1/2}$ in prelim. loop / 3 now $dy/dt_{3/2}$ in next loop
29	23	51	07	STO + 7	Δt			
30	24	01		RCL 1	dx/dt	Δt		7 now t_1 in prelim. loop / 7 now t_2 in next loop
31	61			x	$\Delta t\, dx/dt$			
32	23	51	00	STO + 0	$\Delta t\, dx/dt$			0 now x_1 in prelim. loop / 0 now x_2 in next loop
33	24	00		RCL 0	x			
34	14	74	(or 74)	f PAUSE (or R/S)	x			
35	24	04		RCL 4	Δt			
36	24	03		RCL 3	dy/dt	Δt		
37	61			x	$\Delta t\, dy/dt$			
38	23	51	02	STO + 2	$\Delta t\, dy/dt$			
39	24	02		RCL 2	y			2 now y_1 in prelim. loop / 2 now y_2 in next loop
40	14	74	(or 74)	f PAUSE (or R/S)	y			
41	13	03		GTO 03	y			To new loop
42	22			R ↓	$-C_x$	$-C_y$		
43	02			2	2	$-C_x$	$-C_y$	
44	71			÷	$-C_x/2$	$-C_y$		
45	21			$x \gtrless y$	$-C_y$	$-C_x/2$		
46	02			2	2	$-C_y$		
47	71			÷	$-C_y/2$	$-C_x/2$		
48	21			$x \gtrless y$	$-C_x/2$	$-C_y/2$	$-C_x/2$	
49	13	21		GTO 21	$-C_x/2$	$-C_x/2$		To continue loop

initial $x = 1$, initial $dx/dt = 0$, initial $y = 0$,
initial $dy/dt = 0.9$, $\Delta t = 0.1$, $\alpha = 1$, $\beta = -1.5$

initial $x = 1$, initial $dx/dt = 0$, initial $y = 0$,
initial $dy/dt = 0.8$, $\Delta t = 0.1$, $\alpha = 1$, $\beta = -1.5$

The first set of conditions produces the circular orbit plotted, along with the others, in Fig. 6-3. The dots give the location of the orbiting body at equally spaced time intervals. Thus the figure has the properties of a strobe photo, and you can get a qualitative idea of the speed at any point in the orbit as it is measured by the distance to the next point. Quantitative values for dx/dt and dy/dt can be obtained for any point while running the program by recalling the contents of the registers holding these quantities when the calculator is stopped to display x or y. Observe that the speed is constant everywhere in the circular orbit. The time required to complete one orbit, or the *orbital period T*, can be measured by stopping the calculator at the end of the first orbit and recalling the contents of the register holding the time t. For the circular orbit the value is $T \lesssim 6.3$, where the notation means that the point for $t = 6.3$ was a little way into the next

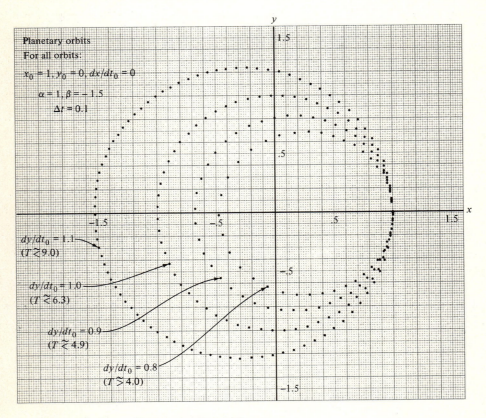

Figure 6-3. A set of bound planetary orbits.

orbit so that 6.3 is a slight overestimate. If a more accurate value is obtained by using the plotted location of that point and the point for $t = 6.2$ to interpolate, the result is $T \simeq 6.28$. Of course, this is really 2π, as it should be for a body moving with unit speed around a circle of unit radius.

In the next example the initial value of dy/dt is 10 percent higher than the value leading to a circular orbit. The resulting orbit is no longer circular, as you can verify by using a ruler to compare its "height" with its "width," or by reading these dimensions from the x and y scales of the plot. At the point in the orbit opposite to the initial point, the body is considerably farther from the fixed body at the center of the coordinates. Also, the orbital speed is no longer constant. As the spacing of adjacent points shows, the speed is lowest when the orbiting body is at its greatest distance from the fixed body. The period for the orbit is increased to the value $T \gtrsim 9.0$.

In the third example a noncircular orbit was obtained by making a 10 percent reduction in the initial value of dy/dt compared to the value leading to a circular orbit. For this case the orbiting body gets closer to the fixed body at the point opposite its initial point, and there its speed is higher than its initial speed. The orbital period is reduced to $T \lesssim 4.9$.

The final example involves an additional 10 percent reduction in the initial value of dy/dt. The results are like the preceding one, but exaggerated. The orbital period is $T \gtrsim 4.0$; that is, the point for $t = 4.0$ is just a little before the end of the first orbit, so 4.0 is a slight underestimate.   ////

A circular orbit is obtained when two conditions are satisfied: (1) the initial motion must be directed perpendicular to the line connecting the initial position to the center, for reasons that are immediately apparent if you consider the geometry of a circle, and (2) the initial speed must have precisely the value for which the centripetal acceleration in a circular orbit (the acceleration towards the center of the circle), of radius equal to the initial distance to the center, equals the force acting at that distance divided by the mass of the body.

When the initial speed is somewhat higher, the force is not strong enough at that distance to bend the trajectory into a circle because the initial linear momentum of the body (its mass times its velocity) is too large. So the body starts off on a trajectory which lies outside the circular orbit. When the initial speed is somewhat lower than the value leading to a circular orbit, its initial linear momentum is reduced and so the central force is more easily able to change the direction of motion, bending the trajectory into an orbit inside the circular orbit.

The reason why the speed is not constant in a noncircular orbit has to do with the fact that the force acting on the orbiting body is always directed to the center. Thus it cannot exert a torque (twist) on it, and the angular momentum of the body (the component of its linear momentum perpendicular to a line extending to the center times the length of the line) must remain constant. For this to happen, the body must have a higher speed when it is closer to the fixed body at the center and a lower speed when it is farther from that body.

For an orbit lying outside the circular orbit, there is an increase in the total distance covered, and also a decrease in the average speed, so the time T required to traverse an orbit is larger than for a circular orbit. These two effects are both reversed for an orbit lying inside the circular one, leading thereby to a reduction in the orbital period.

The quantitative aspects of the behavior discussed qualitatively in the preceding paragraphs is stated succinctly in *Kepler's three laws of planetary motion*. Be sure to look at Exercises 5-2 through 5-6 which give these laws and indicate how you can accurately verify them by using results of running the program.

Example

initial $x = 1$, initial $dx/dt = 0$, initial $y = 0$,
initial $dy/dt = 1.5$, $\Delta t = 0.1$, $\alpha = 1$, $\beta = -1.5$

Here the body is given an initial speed which is 50 percent higher than the value leading to a circular orbit. As a consequence, its initial linear momentum is so high that the central force cannot bend the trajectory into a bound orbit. Figure 6-4 shows that it escapes on a path approaching a straight line and with a speed that approaches a constant value as it gets farther from the central body. This happens because the force acting between the two bodies decreases rapidly as their separation increases, so that before long the moving body has effectively escaped the influence of that force. ////

Example

initial $x = 1$, initial $dx/dt = 0$, initial $y = 0$
initial $dy/dt = 1.1$, $\Delta t = 0.1$, $\alpha = 1$, $\beta = -(2.1 + 1)/2 = -1.55$

In this example the versatility of the numerical method for solving differential equations is used to study the behavior of a noncircular orbit when the central force deviates slightly from the inverse-square law of Eq. (6-5). Inspection of Eqs. (6-5), (6-7), and (6-8) will show you that for the inverse-square law $\beta = -(2 + 1)/2 = -1.5$, where the 2 in the numerator is the power in the force law. For an inverse 2.1 power force law, the value will be $\beta = -(2.1 + 1)/2 = -1.55$, which is the one used in the example.

In contrast to all the bound orbits investigated with the inverse-square law, Fig. 6-5 shows this orbit does not close on itself. If you use a little care in following the points around two consecutive orbits, you will see that the location on the orbit that is farthest from the center advances, i.e., it rotates from one orbit to the next in the same direction as the rotation within an orbit.

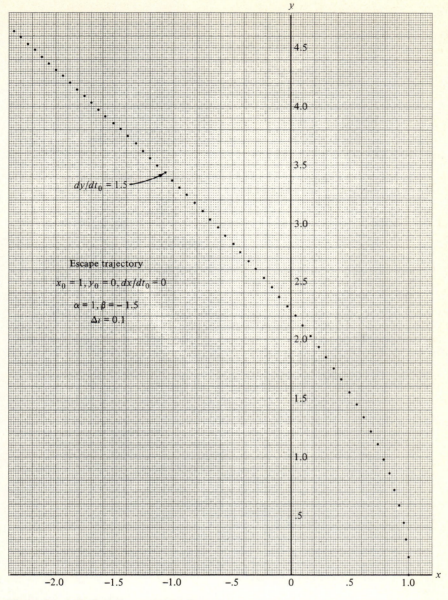

Figure 6-4. An unbound or escape trajectory.

This phenomenon is called *precession*. On a *very* much reduced scale, it is actually found in planetary orbits. For example, the orbit of the planet Mercury precesses by about 0.1° per century. Most of this is due to inverse-square gravitational forces exerted on it by other planets. But

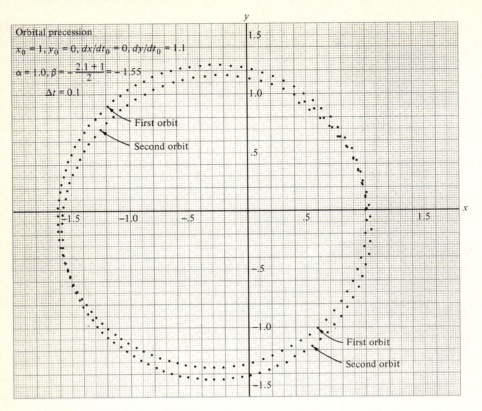

Figure 6-5. Orbital precession in a noninverse-square force.

about one-tenth of the total observed precession, that is, 0.012° per hundred years, is due to a slight departure from the inverse-square law in the gravitational force acting between the sun and Mercury, which arises from a relativistic warping of space in the vicinity of the sun caused by its very large mass.   ////

6-4. α-PARTICLE SCATTERING

You can use the program to study the behavior of a body under the influence of a repulsive force by simply reversing the sign of α.

Example
initial $x = -3$, initial $dx/dt = 2$, initial $y = 0.2$,
initial $dy/dt = 0$, $\Delta t = 0.1$, $\alpha = -1$, $\beta = -1.5$

The results are shown in Fig. 6-6. They represent a body moving from the left in the general direction of the origin of coordinates, where another

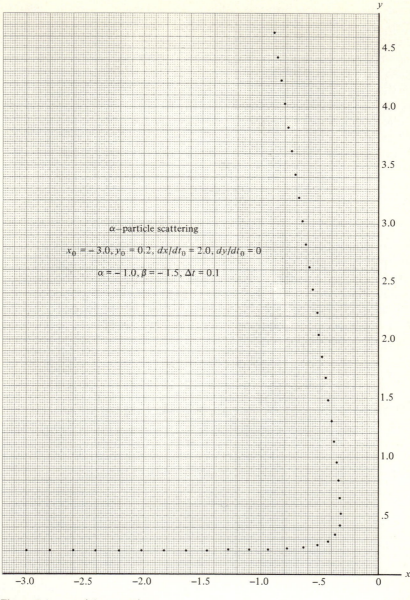

Figure 6-6. α-particle scattering.

much more massive body can be assumed to have a fixed location. The force exerted between the two obeys a repulsive inverse-square law. Thus the example can represent an α particle making a close collision with a nucleus, the force between them being the Coulomb repulsion between their positive charges, providing they do not actually begin to overlap (Ref. 4). The stroboscopic properties of the plotted points make what happens during the scattering very apparent. The α particle moves to the right with almost constant speed at first because it is far from the nucleus. As it approaches the nucleus it is slowed by the repulsion it begins to feel, scattered through a large angle when it gets quite close, and then recedes away regaining the speed it had lost.

If the initial value of y is decreased, will the scattering angle decrease or increase? Use your intuition to make a prediction and then check it by running the program. ////

EXERCISES

6-1. Using the same values $\alpha = 1$, $\beta = -1.5$ employed in the example, run the program to find initial values for x, dx/dt, y, dy/dt which produce a circular orbit of radius 2. It is easier and more accurate than plotting to determine if an orbit is circular by comparing the value displayed for y at $x = 0$ with the value displayed for x at $y = 0$. Measure the period of the orbit by stopping the calculator at the end of two successive revolutions and recording the values of t.

6-2. Kepler's third law of planetary motion, for the case of circular orbits, states that the square of the period of the orbit is proportional to the cube of its radius. Use the periods for the circular orbits of Exercise 6-1 and of the example to test the law.

6-3. The law for the more general case of elliptical orbits states that the square of the period of the orbit is proportional to the cube of half its major axis. Test this form of the law by using data from the four orbits plotted in the example.

6-4. Kepler's first law states that each planet moves in an elliptical orbit with the sun located at one focus of the ellipse. Run the program with initial conditions that will produce a pronounced ellipse, say $\alpha = 1$, $\beta = -1.5$, and initially $x = 1$, $dx/dt = y = 0$, $dy/dt = 0.75$. One of the foci of the orbit is at the origin, of course, and the other can be located by e-voking the obvious symmetry. Use a divider or a ruler to show that the orbit is actually an ellipse because it satisfies the condition that the sum of the distances from any point on it to the two foci is a constant.

6-5. Kepler's second law states that a straight line joining a planet and the sun sweeps out equal areas in space in equal time intervals. Employing the orbit generated in 6-4, verify this law for several typical sets of points on the orbit. Use the fact that the calculator displays x and y at uniformly increasing values of t, and use as a measure of the area of the long thin triangles simply their length times their maximum width.

6-6. Take the points on the orbit of 6-4 for which the velocity is essentially perpendicular to the line joining the point to the origin and run the program again, stopping at these points to record accurately the velocities and distances to the origin at each. Calculate the velocity-times-distance products for each. Explain the relation of these numbers to Kepler's second law, and also to the law of angular-momentum conservation.

6-7. Using the parameters $\alpha = 1$, $\beta = -1.5$ of the example, and also the initial conditions $x = 1$, $dx/dt = y = 0$, find the largest value of dy/dt which will still lead to an elliptical bound orbit instead of a parabolic or hyperbolic escape orbit. Then evaluate the initial values of the kinetic energy $K = m(dy/dt)^2/2$, potential energy $V = -\alpha m/x$, and total energy $E = K + V$. What is the value of E at other points on the orbit? What are the values of E for bound orbits and for escape orbits?

6-8. The orbit for the precise value of total energy that separates bound orbits from escape orbits is actually a parabola. Repeat 6-7 with a very small increase in the initial dy/dt. Plot the orbit that results and compare it with the hyperbolic orbit plotted in the example. Can you distinguish a difference between the two visually? Can you devise a geometrical test along the lines used in 6-4?

6-9. Find a set of initial conditions that, when run in the program, will produce a circular orbit of radius 1, using the parameters $\alpha = 2$, $\beta = -1.5$ to represent doubling the mass of the star located at the origin. Measure the orbital period by stopping the calculator to see t at the beginning and end of a revolution. By comparison with the circular orbit in the example, determine how the period of a circular orbit of a given radius varies with the mass at the origin. The same relation applies to the mass of a planet and the period of its satellite and is used to determine the mass of the planet.

6-10. Use the values of the gravitational constant $G = 6.67 \times 10^{-11}$ $N \cdot m^2/kg^2$ (newton-meters squared per square kilogram), and of the solar mass $M = 1.99 \times 10^{30}$ kg, to evaluate α for the solar system. Then generate a circular orbit of radius 1.50×10^{11} m, the radius of the earth's almost circular orbit about the sun. Determine the value of the period of the orbit and compare it with the length of the year.

6-11. According to Bohr, the electron in the hydrogen atom moves in a circular orbit about the nuclear proton with angular momentum $h/2\pi = 1.06 \times 10^{-34}$ J·s (joule-seconds). The electron's charge and mass are $q = -1.60 \times 10^{-19}$ C (coulomb) and $m = 9.11 \times 10^{-31}$ kg, while the proton's charge is $Q = +1.60 \times 10^{-19}$ C. Evaluate $\alpha = qQ/4\pi\epsilon_0 m$, where $1/4\pi\epsilon_0 = 8.99 \times 10^9$ N·m²/C². Then use the program to find a circular orbit with a radius and velocity that will give the proper angular momentum. Do this by making an initial guess at the radius, finding a circular orbit by inspecting the x and y displays, calculating the angular momentum, modifying the initially guessed radius as is appropriate, and repeating for one or two more iterations. The radius finally obtained will be Bohr's value of the radius of the hydrogen atom, which is a good approximation to the "average" radius predicted by the more accurate Schroedinger equation.

6-12. Investigate the stability of circular orbits for inverse-square and inverse-cube force laws as follows. Set up a circular inverse-square-law orbit by running the program with $\alpha = 1$, $\beta = -1.5$, and initially $x = 1$, $dx/dt = y = 0$, $dy/dt = 1$. Go through approximately one quarter of an orbit, stop the calculator, and compare the value of y with the initial value of x to verify that the orbit is circular. Then represent giving the planet in orbit a radially outward impulse, due to being hit by a very large meteorite, by adding 0.1 to the contents of the dy/dt register. Restart and watch the values of x at $y = 0$ and of y at $x = 0$. How would you describe the trajectory of the planet after being hit? Next show that a circular orbit is possible with an inverse-cube force law by repeating the procedure for the first quarter of an orbit, except using $\beta = -2.0$. Why is it that if $\alpha = 1$, $\beta = -1.5$ allows a circular orbit of radius 1, then $\alpha = 1$, $\beta = -2.0$ should also allow a circular orbit of that radius? Now again give the planet the same radially outward impulse and watch the values of x at $y = 0$ and of y at $x = 0$. What is the trajectory like in this case? Would it be possible for the sun to have a planetary system if gravity obeyed an inverse-cube law?

6-13. Run the program and plot a number of α-particle scattering trajectories, using the same parameters $\alpha = -1$, $\beta = -1.5$ and initial conditions $x = -3$, $dx/dt = 2$, $dy/dt = 0$, with a number of different initial values of y ranging from 0.02 to 2.00.

6-14. Repeat 6-13 using $\alpha = -2$ to represent doubling the nuclear charge.

6-15. Run the program for α-particle scattering trajectories with the same parameters as in 6-13, but with the initial $y = 0$ to generate a head-on collision and a subsequent backscattering. Record the distance of closest approach of the α particle to the nucleus at the origin. Then evaluate the kinetic energy $K = m(dx/dt)^2/2$, potential energy $V = -\alpha m/x$, and total energy $E = K + V$ initially and at the point of closest approach, and comment on the relation between these quantities.

6-16. When α particles of kinetic energy $K = m(dx/dt)^2/2 = 40$ MeV (megaelectronvolts) $= 6.4 \times 10^{-12}$ J are scattered from uranium nuclei at angles larger than about 60°, departures begin to be apparent from the behavior predicted by making the assumption that the only force acting between the two bodies is their Coulomb force. This is interpreted as due to the onset of the nuclear force arising from overlap of the α particle and the nucleus. Use these experimental observations and the program to estimate the radius of the uranium nucleus in the following way. Evaluate $\alpha = -qQ/4\pi\epsilon_0 m$, where the α particle charge is $q = 2 \times 1.6 \times 10^{-19}$ C, the nuclear charge is $Q = 92 \times 1.6 \times 10^{-19}$ C, the α particle mass is $m = 4 \times 1.7 \times 10^{-27}$ kg, and $1/4\pi\epsilon_0 = 9.0 \times 10^9$ N·m²/C². Next use the quoted value of K to evaluate dx/dt. Then make several runs of the program using different values of y until you get a scattering angle of about 60°. The distance of closest approach of the trajectory giving that scattering angle provides the estimate you seek.

6-17. Study the details of the program and explain precisely what happens in each step.

6-18. Find some other system, not necessarily in the field of physics, to which a coupled half-increment scheme similar to the one used in Chaps. 5 and 6 can be applied. Then write a modification of the programs in those chapters to treat the system.

REFERENCES

1. Eisberg, Robert, and Robert Resnick: *Quantum Physics of Atoms, Molecules, Solids, Nuclei, and Particles,* John Wiley & Sons, Inc., New York, 1974, p. 115.

2. Bueche, Frederick J.: *Introduction to Physics for Scientists and Engineers,* 2d ed., McGraw-Hill Book Company, New York, 1975, pp. 164–168.

    Halliday, David, and Robert Resnick: *Fundamentals of Physics,* John Wiley & Sons, Inc., New York, 1974, p. 249.

    Sears, Francis W., and Mark W. Zemansky: *University Physics,* 4th ed., Addison-Wesley Publishing Company, Inc., Reading, Mass., 1970, p. 59.

    Weidner, Richard, and Robert Sells: *Elementary Classical Physics,* Allyn and Bacon, Inc., Boston, 1973, p. 235.

3. Bueche, Frederick J.: *Introduction to Physics for Scientists and Engineers,* 2d ed., McGraw-Hill Book Company, New York, 1975, p. 325.

    Halliday, David, and Robert Resnick: *Fundamentals of Physics,* John Wiley & Sons, Inc., New York, 1974, p. 423.

    Sears, Francis W., and Mark W. Zemansky: *University Physics,* 4th ed., Addison-Wesley Publishing Company, Inc., Reading, Mass., 1970, p. 336.

    Weidner, Richard, and Robert Sells: *Elementary Classical Physics,* Allyn and Bacon, Inc., Boston, 1973, p. 457.

4. Eisberg, Robert, and Robert Resnick: *Quantum Physics of Atoms, Molecules, Solids, Nuclei, and Particles,* John Wiley & Sons, Inc., New York, 1974, p. 99.

    Weidner, Richard, and Robert Sells: *Elementary Classical Physics,* Allyn and Bacon, Inc., Boston, 1973, p. 461.

SEVEN

RANDOM PROCESSES

7-1. INTRODUCTION

This chapter will give you a break from the routine of setting up and solving differential equations. Its purpose is to show you that programmable pocket calculators have other interesting applications in applied mathematical physics. This will be accomplished by using the calculator to simulate and then analyze a simple experiment that has some profound consequences.

The experiment is illustrated in Fig. 7-1. A box is divided down its center by a partition with a small hole which you keep closed by a movable vane while putting a certain number of molecules in the left half and some other number of molecules in the right half. All the molecules are identical (and their density is not so very high that you have to consider interactions between them). Therefore, when you open the vane each molecule has the same chance as any other to be traveling on a path which happens to carry it through the hole. So it is equally likely for any molecule in either side to be the next one to pass through the hole and end up on the other side. In the experiment you monitor the number n_l of molecules on the left, as that number changes each time a molecule passes through the hole in one direction or the other.

It would be quite difficult to perform the experiment with real molecules in a real box. Fortunately, it is not necessary since you can accurately simulate it on the calculator by using a random-number technique based on the following argument.

If at some instant there are n_l molecules on the left and n_r on the right, then since each molecule has the same chance of being the next to go through the hole, the probability is $n_l/(n_l + n_r)$ that the next one to do so will be one on the left. Now consider picking at random a number u_i from a uniformly distributed set of numbers in the range $0 \leq u_i \leq 1$. The probability that the one you get will have a value in the range

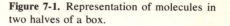

Figure 7-1. Representation of molecules in two halves of a box.

$0 \leq u_i \leq n_l/(n_l + n_r)$ is just equal to the probability that the next passage of a molecule through the hole is from left to right. Do you see why?

Thus the behavior of the system can be simulated in a manner that is in complete agreement with the laws of probability by at each stage making the calculator: (1) generate a random number from 0 to 1; (2) test it against the current value of the fraction of molecules on the left; and (3) "move" a molecule from left to right if the number is smaller than the fraction, or in the opposite direction if it is larger, by subtracting or adding 1 to the appropriate storage registers.

The results of the simulation will lead you to an understanding of why there is an irrevocable tendency for the disorder, or entropy, of the system to increase as time evolves, providing it contains a reasonably large number of molecules. And you will also be able to connect this with the fact that on the macroscopic scale (e.g., where systems contain very large numbers of molecules), the direction in which time actually does evolve is always apparent, even though on the microscopic scale (where the numbers are very small) there is no way to tell the direction of the arrow of time. But first you must learn how to make your calculator produce random numbers.

7-2. RANDOM NUMBER GENERATION

A convenient way to generate a sequence of *quasi random numbers* u_i which are uniformly distributed in the range $0 \leq u_i \leq 1$ is to start with any "seed" number u_0 in the same range and then generate the first number u_1 by calculating

$$u_1 = \text{fractional part of } [(\pi + u_0)^5]$$

The fractional part of a number is what remains after everything to the left of the decimal point has been deleted. The number u_1 is used for whatever purpose you wish and is also used to generate the next number u_2 of the sequence by calculating

$$u_2 = \text{fractional part of } [(\pi + u_1)^5]$$

The scheme can be continued almost indefinitely to yield numbers with the desired properties. Because of the procedure used, a sequence will

Table 7-1. (HP-25) Random-number generator

	0	1	2	3	4	5	6	7
Register Contents:								
	$\underbrace{u_0, u_i}_{\text{preloaded}}$	—	—	—	—	—	—	—

Program

Step	Code		Key Entry	X	Y	Z	T	Comments
00								
01	15	73	g π	π				Start loop
02	24	00	RCL 0	u_0	π			
03		51	+	$\pi + u_0$				
04		05	5	5	$\pi + u_0$			
05	14	03	f y^x	$(\pi + u_0)^5$				
06	15	01	g FRAC	u_1				u_1 = frac. part $[(\pi + u_0)^5]$
07	23	00	STO 0	u_1				To generate u_2
08	14	74	f PAUSE	u_1				
	(or	74)	(or R/S)					
09	13	01	GTO 01	u_1				To new loop

126

Table 7-1. (SR-56) Random-number generator

Register Contents:

0	1	2	3	4	5	6	7	8	9
u_0, u_i preloaded	—	—	—	—	—	—	—	—	—

Program

Step	Code	Key Entry	Comments
00	69	2nd π	π, start loop
01	84	+	
02	34	RCL	
03	00	0	u_0
04	94	=	$\pi + u_0$
05	45	y^x	
06	05	5	$(\pi + u_0)^5$
07	94	=	
08	12	INV	
09	29	2nd Int	$u_1 = $ frac. part $[(\pi + u_0)^5]$

Step	Code	Key Entry	Comments
10	33	STO	
11	00	0	To generate u_2
12	59 (or 41)	2nd PAUSE (or R/S)	
13	59 (or 46)	2nd PAUSE (or 2nd NOP)	
14	22	GTO	
15	00	0	
16	00	0	To new loop

ultimately start to repeat itself. But this will only happen when a number generated is rounded off by the internal limitations of the calculator (by the fact that it only retains a limited number of digits) to be equal to a number earlier in the sequence. This is in no way a practical limitation because the two numbers must be equal to within all the digits retained internally by the calculator (in contrast to the generally smaller number of digits that you choose to display).

The programs in Tables 7-1 (HP-25) and (SR-56) carry out this scheme. They are entered and run as follows.

For the HP-25
Switch to **PRGM**; key **f** **PRGM** to clear; enter the program, choosing either the pause or halt instruction in step 08; switch back to **RUN**; set to step 00 by keying **f** **PRGM**. Then choose any seed u_0 in the range 0 to 1 and enter its value on the keyboard, followed by **STO 0**. To start, push **R/S**; to re-start after a program-controlled stop, push **R/S**; to stop when running, push **R/S**. To generate a different sequence you must start with a different value of the seed u_0.

For the SR-56
Key **2nd** **CP** to clear; key **LRN**; enter the program, choosing either the pause-pause or the halt-no operation instructions for steps 12 and 13; key **LRN**; set to step 00 by keying **RST**. Then choose any seed u_0 in the range 0 to 1 and enter its value on the keyboard, followed by **STO 0**. To start, push **R/S**; to restart after a program-controlled stop, push **R/S**; to stop when running, push **R/S**. To generate a different sequence you must start with a different value of the seed u_0.

Examples (HP-25)
$u_0 = 0.261832695$
$u_i = 0.6476, 0.1117, 0.4350, 0.2956, 0.7635, 0.1542, 0.8773, \ldots$
$u_0 = 0.261832696$
$u_i = 0.6476, 0.1117, 0.4350, 0.2956, 0.7635, 0.1542, 0.8773, \ldots$
$u_0 = 0.261832697$
$u_i = 0.6476, 0.1141, 0.7648, 0.6122, 0.3665, 0.2970, 0.7524, \ldots$

Examples (SR-56)
$u_0 = 0.261832695$
$u_i = 0.6476, 0.1116, 0.3816, 0.8532, 0.4077, 0.2980, 0.4004, \ldots$
$u_0 = 0.261832696$
$u_i = 0.6476, 0.1123, 0.7697, 0.3213, 0.9876, 0.4094, 0.6001, \ldots$
$u_0 = 0.261832697$
$u_i = 0.6476, 0.1130, 0.1559, 0.8887, 0.3963, 0.2443, 0.0092, \ldots$

For convenience, all the sequences are shown as they appear when the calculator is set to round off to four decimal places (by keying **f FIX 4** on the HP-25, or **2nd fix 4** on the SR-56). It is necessary to quote separately the three sequences obtained for each calculator, even though the same values of u_0 were used in both, because they are different.   ////

The reason why the two calculators produce different sequences from the same seeds has to do with the fact that one of them retains more digits in the numbers used in its internal calculations than does the other (in contrast to the fact that for both, the maximum number of digits that can be displayed is the same).

If you inspect the sequences obtained in the examples, keeping in mind the method used to obtain them, you should be able to do the following: Explain why the u_i in any sequence are random. Explain why they are, nevertheless, called quasi random numbers. Estimate how many different sequences the calculators are capable of generating. Determine which of the calculators retains a larger number of digits internally.

Less apparent than the randomness of the u_i is the uniformity of their distribution over the range 0 to 1. When you use these random numbers in the program of the next section, you will find evidence that there are as many in the upper half of the range as in the lower half. Exercise 7-8 suggests how you can test their uniformity with finer resolution.

7-3. ENTROPY AND THE ARROW OF TIME

The programs listed in Tables 7-2 (HP-25) or (SR-56) use the random-number-generating programs to perform the simulation of the experiment in the manner explained in detail earlier. They also have a feature not mentioned earlier. Each time a new value is produced for n_l, the number of molecules present in the left half of the box, that value is entered in a permanently programmed statistical routine which is accessed in each calculator through the $\Sigma+$ key. After the "experiment" is finished, it is very easy to read out the average value of n_l, the standard deviation of n_l (a measure of the fluctuations in its value), and the number N of values of n_l that were analyzed to produce these two quantities. To use these programs, proceed with the following instructions.

For the HP-25

Switch to **PRGM,** key **f PRGM** to clear, then enter the program. If you want to record the results, both steps 22 and 28 should be **R/S;** otherwise both should be **f PAUSE.** After it is entered, switch to **RUN,** and set to step 00 by keying **f PRGM.** Key in the value of the seed u_0, followed by **STO 0.** Do the same for the initial number n_l of molecules on the left and n_r of molecules on the right, following the first by **STO 1,** and the second by **STO 2.**

Table 7-2. (HP-25) Entropy and the arrow of time

Register Contents:

0	1	2	3	4	5	6	7
u_0, u_i	n_l	n_r, n	$(\ldots\ldots\ldots$ used by $\Sigma +$; N is in register 3 $\ldots\ldots\ldots)$				

preloaded

Program

Step	Code	Key Entry	X	Y	Z	T	Comments
00							
01	01	0	0				
02	23 03	STO 3	0				3 zeroed
03	23 06	STO 6	0				6 zeroed
04	23 07	STO 7	0				7 zeroed
05	24 01	RCL 1	n_l				
06	23 51 02	STO + 2	n_l				2 now $n = n_l + n_r$
07	15 73	g π	π				Start loop
08	24 00	RCL 0	u_0	π			
09	51	+	$\pi + u_0$				
10	05	5	5	$\pi + u_0$			
11	14 03	f y^x	$(\pi + u_0)^5$				
12	15 01	g FRAC	u_1				$u_1 = $ frac. part $[(\pi + u_0)^5]$
13	23 00	STO 0	u_1				To generate u_2
14	24 01	RCL 1	n_l	u_1			
15	24 02	RCL 2	n	n_l	u_1		

130

16	71	÷	n_l/n	u_1	Test u_1 to determine direction molecule moves
17	14 51	f $x \geqslant y$	n_l/n	u_1	
18	13 25	GTO 25	n_l/n		
19	01	1	1		
20	51 01	STO + 1	1	u_1	1 now $n_l + 1$ if moved to left
21	24 01	RCL 1	$n_l + 1$		
22	14 74 (or 74)	f PAUSE (or R/S)	$n_l + 1$		
23	23 25	Σ+	$n_l + 1$		
24	13 07	GTO 07	$n_l + 1$		To new loop
25	01	1	1		
26	41 01	STO − 1	1		1 now $n_l - 1$ if moved to right
27	24 01	RCL 1	$n_l - 1$		
28	14 74 (or 74)	f PAUSE (or R/S)	$n_l - 1$		
29	23 25	Σ+	$n_l - 1$		
30	13 07	GTO 07	$n_l - 1$		To new loop

Table 7-2. (SR-56) Entropy and the arrow of time

Register Contents:

0	1	2	3	4	5	6	7	8	9
u_0, u_i	n_l	n_r, n	—	—	(used by $\Sigma+$; N is in register 7)			—	—
preloaded									

Program

Step	Code	Key Entry	Comments
00	00	**0**	0
01	33	**STO**	
02	05	**5**	5 zeroed
03	33	**STO**	
04	06	**6**	6 zeroed
05	33	**STO**	
06	07	**7**	7 zeroed
07	34	**RCL**	
08	01	**1**	n_l
09	35	**SUM**	
10	02	**2**	2 now $n = n_l + n_r$
11	69	**2nd** π	π, start loop
12	84	**+**	
13	34	**RCL**	
14	00	**0**	u_0
15	94	**=**	$\pi + u_0$

Step	Code	Key Entry	Comments
31	04	**4**	
32	05	**5**	
33	01	**1**	1
34	35	**SUM**	
35	01	**1**	
36	34	**RCL**	1 now n_l + 1 if moved to left
37	01	**1**	
38	59	**2nd PAUSE**	
	(or 41)	**(or R/S)**	
39	59	**2nd PAUSE**	
	(or 46)	**(or 2nd NOP)**	
40	26	**2nd f(n)**	
41	04	$\Sigma+$	
42	22	**GTO**	
43	01	**1**	
44	01	**1**	To new loop

Line	Code	Key	Description
16	45	y^x	
17	05	5	
18	94	=	$(\pi + u_0)^5$
19	12	INV	
20	29	2nd Int	u_1 = frac. part $[(\pi + u_0)^5]$
21	33	STO	To generate u_2
22	00	0	Test register loaded with u_1
23	32	$x \leq t$	
24	34	RCL	
25	01	1	n_l
26	54	÷	
27	34	RCL	
28	02	2	n
29	94	=	n_l/n
30	47	2nd $x \geq t$	Test u_1 to determine direction molecule moves

Line	Code	Key	Description
45	01	1	1 now $n_l - 1$ if moved to right
46	12	INV	
47	35	SUM	
48	01	1	
49	34	RCL	
50	01	1	
51	59 (or 41)	2nd PAUSE (or R/S)	
52	59 (or 46)	2nd PAUSE (or 2nd NOP)	
53	26	2nd f(n)	
54	04	Σ+	
55	22	GTO	
56	01	1	
57	01	1	To new loop

Key f **FIX 0,** for the most convenient display of the integral number n_l. Start by pushing **R/S,** and stop by doing the same. After a run, key f **FIX 3,** obtain the standard deviation of n_l by keying f **s,** its average by keying f **x̄,** and then push the ÷ key to calculate the ratio of the standard deviation to the mean, which is a measure of the fluctuations per molecule. To obtain the number N of values of n_l that were analyzed statistically, key **RCL 3.** It is possible to zero the Σ+ key registers during a run, but this is a cumbersome procedure that is best avoided.

For the SR-56

Key **2nd CP** to clear, key **LRN,** then enter the program. If you want to record the results, both steps 38, 39 and steps 51, 52 should be **R/S, 2nd NOP;** otherwise both should be **2nd pause, 2nd pause.** After it is entered, key **LRN,** and set to step 00 by keying **RST.** Key in the value of the seed u_0, followed by **STO 0.** Do the same for the initial number n_l of molecules on the left and n_r of molecules on the right, following the first by **STO 1,** and the second by **STO 2.** Key **2nd fix 0,** for the most convenient display of the integral number n_l. Start by pushing **R/S,** and stop by doing the same. After a run, key **2nd fix 3,** obtain the average of n_l by keying **2nd f(n) Mean,** record it, obtain the standard deviation by keying **2nd f(n) S.Dev,** then calculate the standard deviation divided by the average to obtain a measure of the fluctuations per molecule. To obtain the number N of values of n_l that were analyzed statistically, key **RCL 7.** It is possible to zero the Σ+ key registers during a run, but this is a cumbersome procedure that is best avoided.

Example

initial $n_l = 60$, initial $n_r = 0$

Successive values of n_l are shown in the lower part of Fig. 7-2. The seed u_0 which produced these values is not quoted because it influences only the details of the behavior of the experiment. The general features are the same for all u_0.

When the experiment began, n_l began to decrease as it obviously must. The decrease was uniform until there were enough molecules on the right side of the box that one of them happened to be the one to be chosen by chance to go through the hole. At this point there was a small upward fluctuation superimposed on the continuing downward trend in n_l. The closer n_l got to 30 (one-half the total number of molecules in the box), the more pronounced its fluctuations. It ended up fluctuating around an average value of about 30.

To determine this value more accurately, a statistical analysis of the n_l values was made over an interval starting at the point indicated by the small arrow and continuing until 128 molecules passed through the hole. The average of n_l was found to be 30.156. Since n_l fluctuates, you would not expect its average over a restricted number of values to be precisely

Entropy and the arrow of time

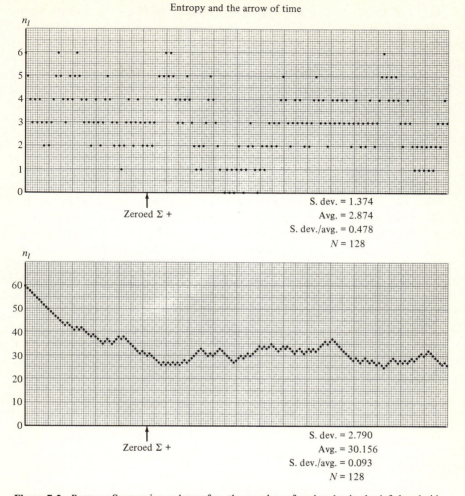

Figure 7-2. *Bottom:* Successive values of n_l, the number of molecules in the left-hand side of the box, in an experiment in which n_l was initially 60 and n_r, the number in the right-hand side, was initially 0. *Top:* The same, except with initial values for n_l and n_r of 6 and 0, respectively.

30. A measure of its fluctuations is given by the standard deviation of n_l, which was found to be 2.790. Also of interest is the standard deviation divided by the average, which was 0.093. $////$

Example

initial $n_l = 6$, initial $n_r = 0$

The values assumed by n_l in this example are shown in the upper part of Fig. 7-2. Here there is also a decrease in n_l at the very beginning of the experiment. There has to be since the first molecule to go through the hole certainly moves from left to right. But that is about all that can be said about the presence of a trend in this example. In fact, the behavior of n_l is completely dominated by its fluctuations. Note, for instance, that on several occasions during the experiment all six molecules fluctuated back into the left side of the box.

For comparison with the preceding example, a statistical analysis was made on the n_l values over the same interval. The average was 2.874, the standard deviation was 1.374, and the standard deviation divided by the average was 0.478.

The average of n_l was almost as far from 3 (one-half the total number of molecules) as the average of n_l was from 30 in the experiment carried out in the previous example. This is not surprising since the standard deviation of n_l was within a factor of 2 of being as large as in that example. Actually, a direct comparison of the standard deviations is a little misleading, as it fails to convey the obvious fact that the fluctuations in this example are very much more significant. The standard deviation divided by the average is more meaningful since it measures the fluctuations per molecule. It was about five times larger in this experiment.   ////

What does all this have to do with entropy and the arrow of time? Entropy is a quantity that arises in two fields of physics—thermodynamics and statistical mechanics. Definitions of entropy which are apparently different, but actually equivalent, are used in each field (Ref. 1). The one that is pertinent here, expressed in qualitative terms, is: "*Entropy* is a measure of the extent of the disorder in a system."

Consider the first experiment. It started off with all 60 molecules in the left side of the box. In this state the system can be said to have a high degree of order, or a low degree of disorder, defining thereby the meaning of these two words as used in this context. Thus the initial entropy of the system was low. After the experiment ran for a while, the molecules distributed themselves through both halves of the box. This happened spontaneously—the system was completely isolated from external influences once you started the experiment by opening the hole. Since the system lost its initial ordering and became disordered, its entropy became high.

The experiment provides and example of one form of the second law of thermodynamics (Ref. 2):

> In a natural process the entropy of an isolated system containing many particles will spontaneously increase.

The basis of the law is that a state of disorder, or high entropy, is the most probable one because there are many more ways to achieve disorder than to achieve order.

How many particles must the system contain for the second law to apply? The experiment just discussed shows you that it applies quite well with 60. Though there continue to be fluctuations in the number of molecules in the left side of the box after that number works its way to approximately 30, it is apparent that you could wait your lifetime without seeing the giant fluctuation that would make most of the molecules temporarily pile into the left side, significantly decreasing the entropy of the system. If the box contained as many gas molecules as in a liter at atmospheric pressure and room temperature, that is, $\sim 10^{20}$, in the lifetime of the solar system such a fluctuation probably would not occur.

On the other hand, the behavior of the experiment with only 6 molecules in the box bore no relation to the predictions of the second law of thermodynamics. The fluctuations were so large, because the number of molecules was so small, that they completely overrode any general trend.

As for the arrow of time, what is involved is the question of how nature defines the natural direction of the flight of time. To understand the question you must understand that the basic laws of physics describing the behavior of individual particles apparently make no distinction as to the direction of time. For instance, Newton's law $d^2x/dt^2 = F/m$ is unchanged if t is replaced by $-t$, since $d^2x/d(-t)^2 = d^2x/dt^2$. You can see this if you look again at Fig. 6-6. If you have not recently seen the figure, you may not remember whether the α particle was incident on the nucleus from the top of the figure moving down, or incident from the left of the figure moving right. In fact, either is a possible motion of the α particle. Therefore, if you saw a motion picture of the process with the α particle incident from the top moving down, you would not be able to tell whether that was what actually happened or whether you were being shown a movie of the α particle incident from the left moving right, but with the direction of time reversed because the film was being run backwards.

Another example is found in the top part of Fig. 7-2. If you were presented with this sequence of values of n_l for the system of 6 molecules, you could not use it to distinguish the direction of the arrow of time. It could just as well represent the results of an experiment in which the initial value of n_l was 3, plotted with that point on the right side of the figure and each subsequent value plotted one additional unit of distance to the left.

But if you look at the bottom part of Fig. 7-2, you can tell immediately that it is plotted with the natural direction of time increasing to the right. If someone showed you a movie of the molecules in the box in which the initial value of n_l was 26, with n_l subsequently fluctuating around 30 for a minute or two and then spontaneously building up to a value of 60, you would *know* that the film was being run backwards. In its own way, nature has contrived to define the direction of the arrow of time by the behavior of many-particle systems, even though it is not defined by the behavior of the few-particle systems which are their constituent parts.

Reference 2 contains an extensive discussion of the subject of this chapter.

EXERCISES

7-1. Explain precisely why the random-number-generating scheme should, in fact, generate a sequence of numbers in which each has no apparent relation to the other, and which are uniformly distributed from 0 to 1.

7-2. Do a number of additional runs of the molecules-in-the-box program using the same initial values of $n_l = 60$, $n_r = 0$, but a different u_0 each time. Plot one or two, and just watch the display for the others. In what ways are the results of the individual experiments different from each other, and in what ways are they the same?

7-3. Run the program using the initial values $n_l = 0$, $n_r = 60$ and discuss the results.

7-4. Make a run with $n_l = 30$, $n_r = 30$ initially, stopping it periodically to record the standard deviation and average of n_l, as well as the number N of values of n_l that have been analyzed. Then plot the standard deviation, the average, and the standard deviation divided by the average, versus N. Comment on your results.

7-5. Run the program for a few cases with $n_l = 6$, $n_r = 0$ and with $n_l = 0$, $n_r = 6$. Compare the results with Exercises 7-2 and 7-3.

7-6. Run the program starting with equal distributions for the following cases: $n_l = n_r = 2, 4, 8, 16, \ldots$. Stop the calculator the first time n_l equals the initial value of $n_l + n_r$. Record the number N of moves which was required for all the molecules to go, in an extreme fluctuation, to one side of the box. The number of molecules that are contained in one liter of gas at room temperature and pressure exceeds 10^{20}. How difficult is it for an initial $n_l = 10^{20}$, $n_r = 0$ distribution to become a $n_l = 0.5 \times 10^{20}$, $n_r = 0.5 \times 10^{20}$ distribution? How difficult is it for an initial $n_l = 0.5 \times 10^{20}$, $n_r = 0.5 \times 10^{20}$ distribution to become a $n_l = 10^{20}$, $n_r = 0$ distribution?

7-7. Study the details of the programs and explain precisely what happens in each step.

7-8. Write a program to test whether or not the quasi random numbers are uniformly distributed by sorting them into, say 5, equal width bins. Use it on the number-generation program. Then discuss the results you obtain with particular reference to the following question: Since there is only time to run a finite sample through the sorting program, how large can the fluctuations in the contents of each bin be before there is reason to question the uniformity of the distribution?

7-9. Devise your own scheme to produce uniformly distributed quasi random numbers. Write a program to implement it. Inspect the results of a number of runs for evidence of randomness. Then run a sample through the sorting program of 7-8 to test for uniformity.

7-10. Write a program using quasi random numbers to simulate a one-dimensional "random walk." Discuss how it can be used to model Brownian motion, gas diffusion, or multiple small-angle α-particle scattering.

REFERENCES

1. Bueche, Frederick J.: *Introduction to Physics for Scientists and Engineers,* 2d ed., McGraw-Hill Book Company, New York, 1975, pp. 313–320.

   Halliday, David, and Robert Resnick: *Fundamentals of Physics,* John Wiley & Sons, Inc., New York, 1974, p. 415.

   Sears, Francis W., and Mark W. Zemansky: *University Physics,* 4th ed., Addison-Wesley Publishing Company, Inc., Reading, Mass., 1970, pp. 278–279.

   Weidner, Richard, and Robert Sells: *Elementary Classical Physics,* Allyn and Bacon, Inc., Boston, 1973, p. 445.

2. Reif, Frederick: *Statistical Physics,* McGraw-Hill Book Company, New York, 1967, p. 5.

CHAPTER
EIGHT

SCHROEDINGER'S EQUATION

8-1. INTRODUCTION

Schroedinger's equation is to quantum mechanics what Newton's law is to classical mechanics. Like Newton's law, it was obtained by postulates designed to ensure its agreement with a few key phenomena, then later was found to be capable of explaining a very large number of others. Its field of application, quantum mechanics, is the mechanics of microscopic systems—atoms, molecules, solids, nuclei, and elementary particles. For this reason, the physical phenomena to which Schroedinger's equation pertains may be less familiar to you than the macroscopic phenomena, such as those treated in earlier chapters, that fall within the field of classical mechanics. Nevertheless, the properties of quantum-mechanical systems are just as important as the properties of classical-mechanical systems. In this chapter you will see among other things how solutions of Schroedinger's equation explain the most important property—energy quantization.

To obtain solutions of Schroedinger's equation you will return to the main stream of the book and employ the general half-increment method. From the mathematical point of view, it is just another differential equation to which the numerical method can be applied immediately. But there are interesting differences in the way the solutions are used, as you will see.

8-2. PLAUSIBILITY ARGUMENT FOR SCHROEDINGER'S EQUATION

Before solving the equation you must know its form. The purpose of this section is to take you through an argument that starts with three properties it is postulated to have, employs them in a simple calculation, and then makes a final postulate that produces the equation. The argument is

140

not a derivation; Schroedinger's equation cannot be derived from something more basic, any more than Newton's law can be so derived. But the argument can help make the equation seem more plausible to you than it would be if it were simply quoted without any prior consideration. This section will make some minimal demands on your physics background. If you are willing to take Schroedinger's equation itself as the basic postulate in the same way that you take Newton's law as a basic postulate, you can skip to the next section.

Schroedinger's equation is a generalization of *de Broglie's relation* (Ref. 1)

$$mv = \frac{h}{\lambda} \tag{8-1}$$

between the momentum mv of a particle of mass m and constant speed v, the wavelength λ of a wave which is associated with the particle, and *Planck's constant*

$$h = 6.63 \times 10^{-34} \quad \text{J·s} \tag{8-2}$$

It is said that a wave is associated with the particle because many experiments show that in certain circumstances a particle moves as if it were governed by the propagation of an associated wave. Specifically, experiments show there are striking diffraction effects in the behavior of microscopic particles (like electrons) and diffraction can be understood *only* on the basis of the behavior of waves (Ref. 2). These effects are not seen in the motion of macroscopic particles (like billiard balls) because Planck's constant is so small that the *de Broglie wavelength* $\lambda = h/mv$ is far too minute to lead to measurable diffraction, unless the momentum mv is also as small as it is for microscopic particles of extremely low mass m. The first postulate is that Schroedinger's equation be consistent with Eq. (8-1).

The second postulate is related to the first. It is that the function describing the mathematical form of the wave associated with a particle whose de Broglie wavelength is λ must be a sinusoidal, say a sine, with that wavelength. The justification is that a sinusoidal is the simplest oscillatory function for which a unique constant wavelength can be defined. The function can be called the *wavefunction* and is symbolized by ψ. Thus for a particle moving along the x axis with de Broglie wavelength λ, the wavefunction is taken to be

$$\psi = \sin\left(\frac{2\pi x}{\lambda}\right) \tag{8-3}$$

Note that λ is indeed the wavelength of this sinusoidal wave since if x increases by the amount λ, then the argument of the sine increases by 2π, and so the sine will go through one full cycle. Note also that Eq. (8-3) is meaningful only for the case of a particle of constant speed v at all positions x for which Eq. (8-1) says λ will be a constant. It is not consistent to speak of a wavelength that varies with position significantly, as such a concept is not well defined. (If you don't believe this, draw a careful sketch of an oscillatory function in which the oscillations bunch closer and closer together with increasing x. Then try to decide, for a particular x, what the wavelength is.) The second postulate is that Schroedinger's equation must be consistent with Eq. (8-3), i.e., that it have (8-3) as a solution for the case of a particle of constant speed.

The third postulate is the reasonable one that Schroedinger's equation be consistent with the law of energy conservation

$$K + V = E \tag{8-4}$$

where K is the kinetic energy of the particle, V is its potential energy, and E is its total energy.

The first step in the calculation is to combine Eqs. (8-1) and (8-4). You express the kinetic energy of the particle in terms of its mass and speed

$$K = \frac{mv^2}{2}$$

Then (8-4) becomes

$$\frac{mv^2}{2} + V = E \tag{8-5}$$

Using (8-1) this immediately gives

$$\frac{h^2}{2m\lambda^2} = E - V$$

or

$$\frac{1}{\lambda^2} = \frac{2m}{h^2}(E - V) \tag{8-6}$$

You are trying to develop a differential equation which has (8-3) as a solution and is also consistent with (8-6). So the next step is to generate some derivatives. Try taking the second derivative of (8-3) with respect to its independent variable x. You first have

$$\frac{d\psi}{dx} = \frac{2\pi}{\lambda}\cos\left(\frac{2\pi x}{\lambda}\right)$$

Then you have

$$\frac{d^2\psi}{dx^2} = -\left(\frac{2\pi}{\lambda}\right)^2\sin\left(\frac{2\pi x}{\lambda}\right)$$

(If you do not know how to differentiate a sine or cosine, one thing you can do is to use the numerical techniques of Chap. 1 to verify these two equalities.) Using Eq. (8-3) on the right side of the second derivative, it simplifies to

$$\frac{d^2\psi}{dx^2} = -\left(\frac{2\pi}{\lambda}\right)^2\psi$$

Now substitute from (8-6) for $1/\lambda^2$ to obtain

$$\frac{d^2\psi}{dx^2} = -\frac{8\pi^2 m}{h^2}(E - V)\psi \qquad (8\text{-}7)$$

This equation is consistent with (8-1) and (8-4) and clearly has (8-3) for a solution. In fact, it is Schroedinger's equation for a special case. It is a special case because it was obtained by using the wavefunction (8-3) which pertains to a particle of constant speed v, as discussed above. This means that the kinetic energy $mv^2/2$ of the particle is constant and so, according to Eq. (8-5), its potential energy V must also be a constant. Thus in (8-7) the quantity V is a constant.

Now comes the final postulate. It is simply to take Eq. (8-7) as valid even for the case where the potential energy V of the particle is not constant but, instead, is a function of position $V(x)$. Doing this, you have the *Schroedinger equation*

$$\frac{d^2\psi}{dx^2} = -\frac{8\pi^2 m}{h^2}[E - V(x)]\psi \qquad (8\text{-}8)$$

written in a form to which it will be particularly convenient for you to apply the half-increment method prescription for obtaining a numerical solution.

Several things should be said. First, technically correct terminology would be to call Eq. (8-8) the *time-independent Schroedinger equation*. There is also a more complicated time-dependent Schroedinger equation, but it is not needed in an elementary treatment of quantum mechanics. Second, to distinguish between them, a solution to the time-dependent Schroedinger equation is often called a wavefunction, written Ψ, and a solution ψ to the time-independent equation is then called an *eigenfunction*. But that distinction need not be made here, so it is unnecessary to use the less picturesque terminology for ψ. Third, and most important, is to tell you what really has happened. The de Broglie relation $mv = h/\lambda$ was known experimentally to be a quantitatively correct statement concerning the wavelike behavior of a particle with constant speed v. Schroedinger wanted to be able to treat also the case of a particle with variable speed. He could not do this by continuing to use the algebraic equation $mv = h/\lambda$ since, as explained earlier, it becomes meaningless if the speed v (and therefore the wavelength λ) is not constant. But there is no speed v in the equivalent differential equation (8-7); there is only the related quantity V, the potential energy. Therefore, it was *consistent* for him to generalize (8-7) by postulate to obtain (8-8), which is supposed to pertain to the case of a particle moving with varying speed because its potential energy $V(x)$ varies with position. Whether or not it was *correct* is a matter that experiment alone could decide. Suffice it to say that in the intervening fifty years, countless experiments have shown the correctness of Schroedinger's equation.

8-3. HARMONIC OSCILLATOR

The *Schroedinger equation* for a particle of mass m, whose total energy is the constant E and whose potential energy as a function of its position is $V(x)$, is

$$\frac{d^2\psi}{dx^2} = -\frac{8\pi^2 m}{h^2}[E - V(x)]\psi \qquad (8\text{-}8)$$

where h is the value of Planck's constant quoted in Eq. (8-2). It determines the properties of the so-called wavefunctions ψ since ψ, which is a function of x, is a solution to the differential equation for given m, E, and $V(x)$.

The wavefunctions in turn determine the behavior of the particle through a postulate due to Born. According to him, the probability of finding the particle in a small region surrounding the position x is proportional to ψ^2, the square of the wavefunction, evaluated for that x. *Born's postulate* is as basic to quantum mechanics as is Schroedinger's. It originally was justified by analogy to known properties of electromagnetism: At any point the square of the function describing an electromagnetic wave (i.e., the intensity of its electric or magnetic field) is proportional to its energy content in a small region surrounding that point, and this is a measure of the probability of finding an electromagnetic particle (i.e., a photon) in the region. But, as is true for the Schroedinger equation itself, the ultimate justification for Born's postulate is that it subsequently has been found to agree with a very large set of experiments.

You will note that in quantum mechanics the physical properties of the system containing the particle are specified by the potential energy $V(x)$ that the particle has as a result of the force F which the system exerts on it, and not directly in terms of the force. However, F and $V(x)$ have the relation

$$F = -\frac{dV(x)}{dx} \qquad (8\text{-}9)$$

(If you are unfamiliar with this, a simple example is derived in Fig. 8-1.) Thus you can always get $V(x)$ from F, or vice versa. To illustrate the latter, the potential energy $V(x)$ of the particle in the harmonic oscillator of Fig. 4-1 is

$$V(x) = \frac{kx^2}{2} \qquad (8\text{-}10)$$

where k is the constant describing the stiffness of the spring. This can be verified by using Eq. (8-9) to calculate the corresponding force

$$F = -\frac{d}{dx}\left(\frac{kx^2}{2}\right) = -\frac{2kx}{2} = -kx$$

and noting that the result agrees with Eq. (4-1). (If you do not know the

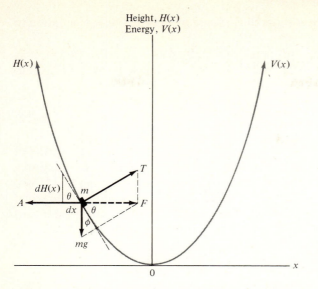

Figure 8-1. Gravitational example of the relation between force and potential energy. A mass m is held in equilibrium on a smooth track by an applied force A canceling the force F, which is the sum of the gravitational force mg and the force T exerted by the track. You can see that $F \cos \theta = mg \cos \phi = mg \cos (90° - \theta) = mg \sin \theta$, or $F = mg \sin \theta / \cos \theta = mg \tan \theta$. Also, $\tan \theta = -dH(x)/dx$, where the sign is minus since it takes a negative change in x to make a positive change in $H(x)$. Thus, $F = -mg \, dH(x)/dx$. But the potential energy of the mass relative to its value at zero height is $V(x) = mg \, H(x)$. So, $dV(x) = d[mg \, H(x)] = mg \, dH(x)$, since mg is a constant. Therefore, $F = -dV(x)/dx$. If the applied force is removed, the mass will move in the direction of positive x under the influence of the force F.

analytical expression for the derivative used to obtain the result, you can always verify it for a few values of x and k by the methods of Chap. 1.) The *harmonic-oscillator potential* $V(x) = kx^2/2$ is plotted in Fig. 8-1.

Schroedinger's equation for the harmonic oscillator is obtained by using Eq. (8-10) in (8-8), yielding

$$\frac{d^2\psi}{dx^2} = -\frac{8\pi^2 m}{h^2}\left[E - \frac{kx^2}{2}\right]\psi \qquad (8\text{-}11)$$

This is the differential equation which quantum mechanics says governs the behavior of the same system that classical mechanics says is governed by $d^2x/dt^2 = -kx/m$. What a difference! Yet in Exercise 8-9 you will find that the two equations lead to equivalent predictions concerning macroscopic oscillators. For microscopic oscillators their predictions have no relation, and experiments show those made by Schroedinger's equation to be the ones that are correct.

Before applying the half-increment method to solve Eq. (8-11), it is worthwhile working it over until it assumes a simpler form. To this end, you can introduce the relation

$$\nu = \frac{1}{2\pi} \sqrt{\frac{k}{m}} \tag{8-12}$$

for the frequency ν at which a particle of mass m at the end of a spring of stiffness k would oscillate, according to classical mechanics. Using this relation does not mean that the results obtained from solving (8-11) will depend on classical mechanics for their validity; actually, you are just using classical mechanics to suggest defining the symbol ν for the convenient grouping of parameters $(1/2\pi) \sqrt{k/m}$. (If you do not know Eq. (8-12), you can verify it by following Exercise 4-3.) In terms of ν, the Schroedinger equation becomes

$$\frac{d^2\psi}{dx^2} = -\left[\frac{8\pi^2 mE}{h^2} - \left(\frac{4\pi^2 m\nu}{h}\right)^2 x^2\right]\psi$$

Defining

$$\beta = \frac{8\pi^2 mE}{h^2} \qquad \alpha = \frac{4\pi^2 m\nu}{h} \tag{8-13}$$

it reduces to the form

$$\frac{d^2\psi}{dx^2} = -(\beta - \alpha^2 x^2)\,\psi \tag{8-14}$$

You can make it look even more attractive by writing it in terms of the dimensionless variable (i.e., the variable is a pure number)

$$u = \sqrt{\alpha}\, x \tag{8-15}$$

Then

$$\frac{d^2\psi}{dx^2} = \frac{d^2\psi}{d(u/\sqrt{\alpha})^2} = \alpha\frac{d^2\psi}{du^2}$$

since α is a constant and the derivative is therefore like any other fraction, as far as it is concerned. Using this in Eq. (8-14) you have

$$\alpha\frac{d^2\psi}{du^2} = -(\beta - \alpha u^2)\,\psi$$

or

$$\frac{d^2\psi}{du^2} = -\left(\frac{\beta}{\alpha} - u^2\right)\psi$$

Finally, writing

$$\epsilon = \frac{\beta}{\alpha} = \frac{8\pi^2 mE}{h^2}\frac{h}{4\pi^2 m\nu}$$

or
$$\epsilon = \frac{2E}{h\nu} \qquad (8\text{-}16)$$

you have reduced the Schroedinger equation to its simplest form

$$\frac{d^2\psi}{du^2} = -(\epsilon - u^2)\,\psi \qquad (8\text{-}17)$$

It is also in its most universal form, since it contains no constants (they have all been hidden in the definitions of ϵ and u). This being the case, it pertains to all situations no matter what the values of k and m. Once its solutions have been obtained, they can be used to predict the behavior of all quantum-mechanical harmonic oscillators.

Since Eq. (8-17) is of the form

$$\frac{d^2y}{dt^2} = C$$

required for application of the general prescription of Sec. 3-6, its solutions are obtained in the now so familiar way. This is essentially a matter of taking any program of, say, Chap. 4, and rewriting it by changing the independent variable to u, the dependent variable to ψ, and also using the coefficient

$$C = -(\epsilon - u^2)\,\psi \qquad (8\text{-}18)$$

Programs are shown in Tables 8-1 (SR-56) or (HP-25). If you inspect them, you will see that they do, however, have two new features. One is a provision for the calculator to display ψ only after every jth calculational loop. This is to allow you to attain a high degree of numerical accuracy in the solution by reducing Δu without the necessity of inspecting or plotting too many closely spaced values of ψ. The other feature is that to input the numerical value of ϵ, you key it into the calculator immediately before you begin a run by pressing the start button. Although it would not have to be done this way in these programs, it is to maintain uniformity of operation with the program in Table 8-2 (HP-25) where there would otherwise be a shortage of storage registers.

More specifically, the programs are run according to the following instructions.

For the SR-56
Key **2nd CP** to clear, key **LRN,** then key in the program using the desired display option in steps 51 and 52. Key **LRN,** and set to step 00 by keying **RST.** Load initial values of ψ and $d\psi/du$ and the values of Δu and j in registers 1, 2, 3, and 4, respectively. Then key in the value of ϵ and press **R/S** to start. To stop, or restart, or restart after a program-controlled stop, push **R/S.** To see the value of u while stopped by the program, key **RCL 7.** To do so while stopped manually, key **2nd EXC 7,** inspect u, then again key **2nd EXC 7.**

Table 8-1. (SR-56) Harmonic oscillator Schroedinger equation

Register Contents:

0	1	2	3	4	5	6	7	8	9
	ψ	$d\psi/du$	Δu	j	ϵ	—	u	$C, C/2$	—

preloaded
start with ϵ in display register

Program

Step	Code	Key Entry	Comments
00	33	STO	
01	05	5	5 now ϵ
02	00	0	0
03	33	STO	
04	07	7	7 zeroed
05	34	RCL	
06	04	4	j
07	33	STO	
08	00	0	0 set to j
09	56	2nd CP	Test register zeroed
10	34	RCL	Start loop
11	07	7	u
12	43	x^2	u^2
13	74	–	
14	34	RCL	
15	05	5	ϵ
16	94	=	$u^2 - \epsilon$
17	64	x	
18	34	RCL	
19	01	1	ψ

Step	Code	Key Entry	Comments
36	34	RCL 3	Δu
37	03		
38	35	SUM 7	7 now u_1 in prelim. loop, u_2 in next
39	07		
40	64	x	
41	34	RCL 2	$d\psi/du_{1/2}$ in prelim. loop, $d\psi/du_{3/2}$ in next
42	02		
43	94	=	$\Delta nd\psi/du_{1/2}$ in prelim. loop, $\Delta nd\psi/du_{3/2}$ in next
44	35	SUM 1	1 now ψ_1 in prelim. loop, ψ_2 in next
45	01		
46	27	2nd dsz	Test i for routing to skip display
47	01	1	
48	00	0	To new loop
49	34	RCL 1	ψ_1 in prelim. loop, ψ_2 in next
50	01		

148

Step	Code	Key	Comment
20	94	=	$C = (u^2 - \epsilon\psi)$
21	33	STO	
22	08	8	8 now C
23	34	RCL	
24	07	7	u
25	37	2nd $x = t$	Test u for routing in preliminary loop
26	06	6	
27	00	0	
28	34	RCL	
29	03	3	
30	64	x	Δu
31	34	RCL	
32	08	8	$C/2$ in prelim. loop, C in next loop
33	94	=	$\Delta u C/2$ in prelim. loop, $\Delta u C$ in next loop
34	35	SUM	
35	02	2	2 now $d\psi/du_{1/2}$ in prelim. loop, $d\psi/du_{3/2}$ in next

Step	Code	Key	Comment
51	59	2nd PAUSE (or R/S)	
52	59 (or 41)	2nd PAUSE (or 2nd NOP)	
53	34 (or 46)	RCL	
54	04	4	j
55	33	STO	
56	00	0	0 now reset to j
57	22	GTO	
58	01	1	
59	00	0	
60	02	2	2
61	12	INV	
62	30	2nd PROD	
63	08	8	8 now $C/2$
64	22	GTO	
65	02	2	
66	08	8	To continue loop

Table 8-1. (HP-25) Harmonic oscillator Schroedinger equation

Register Contents:

0	1	2	3	4	5	6	7
—	ψ	$d\psi/du$	Δu	j	ϵ	i	u

preloaded start with ϵ in X register

Program

Step	Code	Key Entry	X	Y	Z	T	Comments
00							
01	23 05	STO 5	ϵ				5 now ϵ
02	01	1	1				
03	23 06	STO 6	1				6 set to initial value
04	00	0	0				
05	23 07	STO 7	0				7 zeroed Start loop
06	24 05	RCL 5	ϵ				
07	24 07	RCL 7	u	ϵ			
08	15 02	g x^2	u^2	ϵ			
09	41	−	$\epsilon - u^2$				
10	24 01	RCL 1	ψ	$\epsilon - u^2$			$C = -(\epsilon - u^2)\psi$
11	61	x	$-C$				
12	24 07	RCL 7	u	$-C$			
13	15 71	g $x = 0$	u	$-C$			Test u for routing to put 1/2 in preliminary loop
14	13 36	GTO 36	u	$-C$			
15	22	R↓	$-C$	$-C$			
16	24 03	RCL 3	Δu	$-C$			Will be $-C/2$ in prelim. loop
17	61	x	$-C\Delta u$				
18	23 41 02	STO − 2	$-C\Delta u$				2 now $d\psi/du_{1/2}$ in prelim. loop 2 now $d\psi/du_{3/2}$ in next loop
19	24 03	RCL 3	Δu				

Step				Key		Δu	Comment
20	23	51	07	STO + 7		Δu	7 now u_1 in prelim. loop
21		24	02	RCL 2	$d\psi/du$		7 now u_2 in next loop
22			61	x	$\Delta u\,d\psi/du$		
23	23	51	01	STO + 1	$\Delta u\,d\psi/du$		1 now ψ_1 in prelim. loop
24		24	06	RCL 6	i	i	1 now ψ_2 in next loop
25		24	04	RCL 4	j	i	
26		14	71	f $x=y$	j	i	Test i for routing to display ψ
27		13	31	GTO 31	j		
28			01	1	1		
29	23	51	06	STO + 6	1		Increase i
30		13	06	GTO 06	ψ		To new loop
31		24	01	RCL 1	ψ		
32		14	74	f PAUSE			
			(or 74)	(or R/S)			
33			01	1	1		
34		23	06	STO 6	1		6 reset
35		13	06	GTO 06	1		To new loop
36			22	R↓	$-C$		
37			02	2	2	$-C$	
38			71	÷	$-C/2$		
39		13	16	GTO 16	$-C/2$		To continue loop

For the HP-25

Switch to **PRGM**, key **f** **PRGM** to clear, then key in the program using the desired display option at step 32. Switch back to **RUN,** and set to step 00 by keying **f** **PRGM.** Load initial values of ψ and $d\psi/du$ and values of Δu and j in registers 1, 2, 3, and 4, respectively. Then key in the value of ϵ and press **R/S** to start. To stop, or restart, or restart after a program-controlled stop, push **R/S**. To see the value of u while stopped, key **RCL 7.**

Before giving examples, it should be explained how you can intelligently choose an initial u and the corresponding values of ψ and $d\psi/du$. The harmonic-oscillator potential $V(x) = kx^2/2$ is an even function of x and, therefore, also of u. That is, its value for a particular u_i is exactly equal to its value at $-u_i$. The behavior of the particle moving under the influence of the potential can be expected to show the same symmetry as the potential itself. (This is certainly true for a classical harmonic oscillator.) As the behavior is governed by the value of ψ^2, according to Born's postulate, you can see that ψ^2 should be an even function of u. This means that ψ must be either *even* in u, or *odd* in u. The point is that if either

$$\psi(-u_i) = +\psi(u_i)$$
or
$$\psi(-u_i) = -\psi(u_i)$$

then ψ^2 will have the required property

$$\psi^2(-u_i) = +\psi^2(u_i)$$

Because of the symmetry of ψ^2, it is only necessary to carry out the numerical solution of the equation for the positive half of the u axis. (In fact, the programs will automatically start at $u = 0$ and then increment it positively.) And at $u = 0$ the wavefunction ψ must either have $d\psi/du = 0$, or have $\psi = 0$. The first condition gives the even ψ case, while the second gives the odd ψ case, as you can see from Fig. 8-2. The values used for ψ at $u = 0$ for the even case, or $d\psi/du$ at $u = 0$ for the odd case, do not really matter. Because the differential equation is linear in ψ, increasing or decreasing these by some factor will only scale up or down all the values of its solution ψ, without affecting its overall shape or any other vital properties. The first example below considers an even solution by taking $\psi = 1$ and $d\psi/du = 0$ at $u = 0$.

What about the value of the parameter ϵ? You will observe from Eq. (8-16) that it really is a dimensionless measure of the total energy E of the harmonic oscillator. In classical mechanics, any value of E (or ϵ) would be a possible one for the oscillator. You will soon see that this is not true in quantum mechanics; in fact, the game will be to find which values of ϵ are allowed and which are not. But this still leaves you with the question of how to choose a value of ϵ to start studying the solutions of the equation. Since Eq. (8-17) contains no numerical constants, except the 1 that is the unwritten coefficient of each term, you might as well try first $\epsilon = 1$.

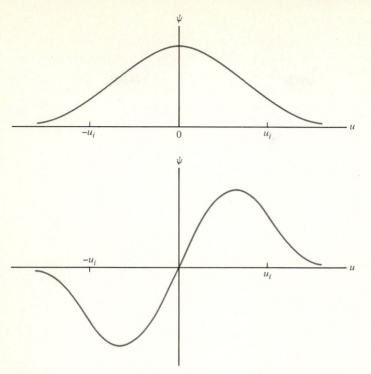

Figure 8-2. *Top:* An even function of u. Its value at any point $-u_i$ equals its value at the corresponding point u_i. Since it has this symmetry about $u = 0$, it is necessary for its slope $d\psi/du$ to be zero at that point if the slope is to be continuous there. *Bottom:* An odd function of u. Its value at any point $-u_i$ equals the negative of its value at the corresponding point u_i. For this case the symmetry requires its value ψ to be zero a $u = 0$, assuming that the function is continuous at that point.

Example

initial $\psi = 1$, initial $d\psi/du = 0$, $\Delta u = 0.05$, $j = 2$, $\epsilon = 1.000$

The results are shown by the crosses labeled $\epsilon = 1.000$ in Fig. 8-3. Note that the curve traced by the crosses starts off from $u = 0$ with a concave downward curvature. This reflects the fact that $d^2\psi/du^2$, which is a measure of the curvature of ψ, is negative for small u where the term $-(\epsilon-u^2)$ of the equation $d^2\psi/du^2 = -(\epsilon - u^2)\psi$ is negative, since ψ itself is positive. At $u = 1$, $\epsilon - u^2 = 0$ since $\epsilon = 1$, so $d^2\psi/du^2$ is zero at that point. This is why the curve is locally straight at $u = 1$. When u exceeds this value, $d^2\psi/du^2$ changes sign since $\epsilon - u^2$ does, and so the curvature becomes concave upward, at least while ψ remains positive. But since the slope $d\psi/dt$ is negative at $u = 1$, the positive $d^2\psi/du^2$ does not prevent ψ from crossing the u axis at about $u = 3.2$. When this happens, $d^2\psi/du^2$

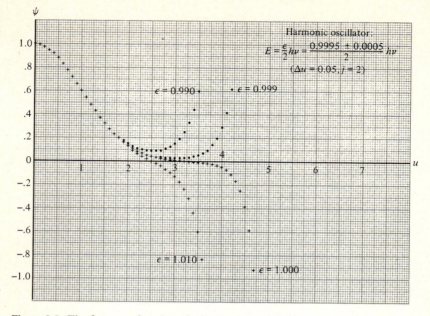

Figure 8-3. The first wavefunction of a harmonic oscillator.

changes sign again because ψ becomes negative, and ψ becomes concave downward. So its negative slope starts to increase and ψ bends rapidly away from the u axis. When this behavior gets started, nothing can prevent it from continuing with ever increasing vigor, and ψ diverges to negative infinity. The reason is that the more negative the value of ψ, the more negative is the value of $d^2\psi/du^2$ because of the relation between them that is imposed by the differential equation $d^2\psi/du^2 = -(\epsilon - u^2)\psi$.

////

The results of this example constitute a solution to the differential equation which is, within the accuracy of the numerical method, a completely correct one for the initial conditions used. It is an acceptable solution to the differential equation from a mathematical point of view, but it is *not* an acceptable wavefunction from a physical point of view. Born's postulate says that ψ^2 is a measure of the probability of finding the associated particle in various locations on the u axis. The behavior of the ψ found in the example will cause the corresponding ψ^2 to increase without limit when u becomes more positive than a not very large number (and when it becomes more negative than minus that number). Thus you must reject this solution because it says the particle has an infinite probability of being everywhere except where it is supposed to be. Where is it supposed to be? Since the potential $V(x)$ has its minimum at $u = 0$, the force acting on the particle is always directed toward that point and the particle should be somewhere in its vicinity. (See Fig. 8-1.)

Example

initial $\psi = 1$, initial $d\psi/du = 0$, $\Delta u = 0.05$, $j = 2$, $\epsilon = 1.010$

In this example an attempt is made to find an acceptable wavefunction by increasing the value of ϵ. Because of the obvious sensitivity of the differential equation, ϵ is increased by only 1 percent. The results are indicated by crosses labeled $\epsilon = 1.010$ in Fig. 8-3. They show that increasing ϵ is the wrong thing to do.   ////

Example

initial $\psi = 1$, initial $d\psi/du = 0$, $\Delta u = 0.05$, $j = 2$, $\epsilon = 0.990$

Here ϵ is decreased, from the value guessed at first, by 1 percent. The resulting behavior of ψ is shown in Fig. 8-3 by the dots labeled $\epsilon = 0.990$, which fuse into the crosses for u smaller than about $u = 1.8$. For this ϵ the concave upward curvature of ψ to the right of $u = 1$ is a little more pronounced than for $\epsilon = 1.000$, since in this region where $u^2 > \epsilon$ the magnitude of the term $\epsilon - u^2$ is larger when ϵ is smaller. But although the curvature of ψ is only slightly larger, its cumulative effect is to succeed in making the slope of ψ go through zero and become positive before ψ itself crosses the axis and becomes negative. When this happens, ψ again starts to diverge to infinity—this time to positive infinity. The farther it gets from the u axis, the larger its rate of change of slope, so the more rapidly it increases.   ////

Example

initial $\psi = 1$, initial $d\psi/dt = 0$, $\Delta u = 0.05$, $j = 2$, $\epsilon = 0.999$

In this example ϵ is made 0.1 percent smaller than in the first example. The results are plotted as dots labeled $\epsilon = 0.999$ in Fig. 8-3. They are similar in character to the preceding example, but the divergence to positive infinity does not occur until u assumes larger values.   ////

By now you can see how it goes. If runs are made with $\epsilon = 1.0000$ and $\epsilon = 0.9999$ using an appreciably smaller value of Δu to reduce the inaccuracy of the numerical method to the lower value dictated by these narrower limits, the ψ found in the former will diverge to negative infinity at a somewhat larger u than for the case shown in the first example, and the ψ found in the latter will diverge to positive infinity at about the same larger u. If this is not apparent, try it.

Will you ever succeed in obtaining, from a numerical solution of the Schroedinger equation, a ψ that never diverges? Not until you have access to a computer of infinite speed that would allow you to use an infinitesimally small Δu, and which also carries an infinite number of digits so as to have zero roundoff error. But is it necessary? What would happen if the ultimate machine were put to work on this problem is apparent enough anyway. The results would look like the points or crosses of the $\epsilon = 0.999$

or $\epsilon = 1.000$ curves to about $u = 3$, and then continue slowly approaching, but never reaching, the u axis. The closer ψ got to zero, the smaller the value of its second derivative $d^2\psi/du^2 = -(\epsilon - u^2)\psi$, and so the smaller its curvature. Thus the differential equation is consistent with an ever closer approach to a straight line lying along the u axis.

If you understand this, you will understand that there is no practical need to go further than the first and fourth examples. These bracket the shape of the allowed wavefunction ψ quite accurately, and bracket the corresponding allowed value of the energy parameter ϵ to within the narrow limits $\epsilon = 0.9995 \pm 0.0005$. To put it another way, the actual energy value for this allowed wavefunction is

$$E = \frac{h\nu}{2} \qquad \text{(within 5 parts in 10,000)}$$

where Eq. (8-16) has been used to go from ϵ to E.

This is not the only value of the energy allowed by quantum mechanics for a harmonic oscillator. See if you can find another allowed ϵ in the range from 0 to below 1. Do not bother to plot ψ; just watch the display when your calculator is running in the pause mode. You will soon conclude that there are no values of ϵ in this range for which the behavior of ψ, as u becomes sufficiently large, is analogous to the behavior displayed in Fig. 8-3. This will be your conclusion whether you search for a ψ which is an even function of u, as in the figure, or an odd function of u. To search for an odd ψ, use the initial conditions $\psi = 0$, $d\psi/du = 1$, at $u = 0$.

Next continue the search to values of ϵ greater than 1. You can speed it up by using a relatively large Δu, reducing the increment for accuracy when you get into a promising range of ϵ. Soon you will find that the next allowed energy occurs at $\epsilon \simeq 3$ and represents a case where ψ is odd in u.

Examples

initial $\psi = 0$, initial $d\psi/du = 1.65$, $\Delta u = 0.02$, $j = 4$, $\epsilon = 2.999$

initial $\psi = 0$, initial $d\psi/du = 1.65$, $\Delta u = 0.02$, $j = 4$, $\epsilon = 3.000$

The initial value of $d\psi/du$ was adjusted after the critical values of ϵ were found by watching the calculator display, and before plotting, so as to make the peak height of ψ for results plotted in Fig. 8-4 be the same as in Fig. 8-3. This facilitates comparison without affecting the shape of the allowed ψ, or the value of the allowed ϵ, since the Schroedinger equation is linear in ψ.

In this case ψ starts with a positive slope from a zero value at $u = 0$. But since its curvature is much higher for small u than for the ψ in Fig. 8-3, it nevertheless is able to bend over before the sign of $\epsilon - u^2$ switches, i.e., before $3 - u^2 = 0$ or $u = \sqrt{3} = 1.73$. For larger u it behaves much as before.

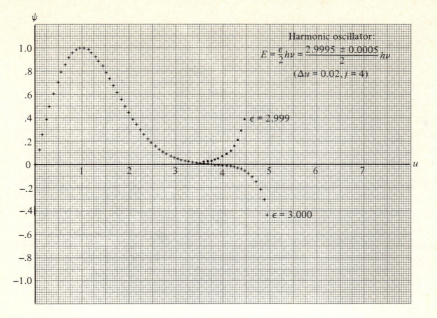

Figure 8-4. The second wavefunction of a harmonic oscillator.

The value of the allowed energy corresponding to the range $\epsilon = 2.9995 \pm 0.0005$ found in this example is

$$E = \frac{3h\nu}{2} \qquad \text{(within 5 parts in 30,000)}$$

////

Run the program to find the next higher allowed value of ϵ (or E) and the corresponding allowed wavefunction ψ. Since the first one is even and the second is odd, what do you think the symmetry of the next will be? Use the appropriate initial conditions. Since the first value of ϵ is 1 and the second is 3, you may also be able to make a good guess about the next value of ϵ. As the curvature of ψ in the region of u where $\epsilon - u^2$ is positive will be even larger for a higher value of ϵ, you can expect that the ψ you seek will oscillate even more rapidly in that region.

The final examples give the ninth allowed solution to the Schroedinger equation for the harmonic oscillator.

Examples

initial $\psi = 1$, initial $d\psi/du = 0$, $\Delta u = 0.01$, $j = 5$ for $u < 5$, $j = 10$ for $u > 5$, $\epsilon = 16.999$

initial $\psi = 1$, initial $d\psi/du = 0$, $\Delta u = 0.01$, $j = 5$ for $u < 5$, $j = 10$ for $u > 5$, $\epsilon = 17.000$

Here the value of the allowed energy is

$$E = \frac{17h\nu}{2} \qquad \text{(within 5 parts in 170,000)}$$

The plots in Fig. 8-5 show that the oscillations in ψ build up until ψ reaches its final peak just before $\epsilon - u^2 = 0$. Exercise 8-9 will give you an understanding of the physical significance of this result. Can you explain its mathematical origin from an argument similar to those above involving the curvature of ψ?   ////

Figure 8-6 summarizes some of the results of this section by showing the harmonic-oscillator potential energy $V(x)$ plus a set of horizontal lines representing the allowed values of total energy E for the oscillator. The vertical scale is energy, either potential or total, and the horizontal scale is position x. Since $V(x) = kx^2/2$, the potential energy plots into a parabola. As any particular E is a constant for all x, it is plotted as a horizontal line.

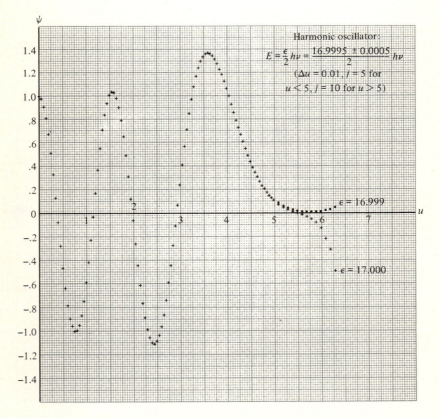

Figure 8-5. The ninth wavefunction of a harmonic oscillator.

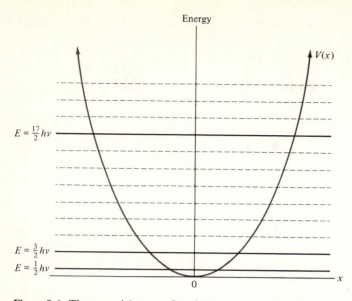

Figure 8-6. The potential energy function $V(x)$ and the first few energy levels E of the harmonic oscillator.

The solid ones are the allowed values of E found in the examples; the dashed ones are other allowed values that you can find by running the program. The entire set of horizontal lines is called the *energy-level diagram* for the harmonic oscillator. Expressed mathematically, the entire set of allowed energies for a harmonic oscillator is

$$E = \frac{h\nu}{2}, \ \frac{3h\nu}{2}, \ \frac{5h\nu}{2}, \ \frac{7h\nu}{2}, \ \dots$$

or

$$E = \left(n + \frac{1}{2}\right) h\nu \qquad n = 0, 1, 2, 3, \dots \qquad (8\text{-}19)$$

where h is Planck's constant and ν is the frequency at which the oscillator would oscillate classically.

For a given energy level, the two intersections of the curve $V(x)$ and the line E determine the range of x within which the particle of a classical oscillator with that energy would be confined, or bound. Inside this region $E > V(x)$, so its kinetic energy is $K = [E - V(x)] > 0$. It would never be found outside where $E < V(x)$ since there $K = [E - V(x)] < 0$, which is impossible in classical mechanics because $K = mv^2/2$; so $K < 0$ means the speed v is an imaginary number. But this is not a strict limitation on the location of the particle in quantum mechanics. The values of x for which $E - V(x) = 0$ correspond to those of u for which $\epsilon - u^2 = 0$. As you have seen in analyzing Figs. 8-3, 8-4, and 8-5, ψ and ψ^2 do extend outside the classical region somewhat. Thus quantum mechanics predicts, and experiment confirms, that there is some probability of the particle being outside. An explanation of how this comes about involves the uncertainty principle and must be left to the references (Ref. 3).

The most important conclusion of this section and chapter can be stated as follows: When the relation between the potential energy V of a particle and its total energy E is such that in classical mechanics the particle would be bound to a limited region of space, then in quantum mechanics only certain discrete values of E are allowed. In these circumstances the total energy of the particle is said to be *quantized*.

8-4. FINITE SQUARE WELL

Quantum mechanics makes interesting predictions about the behavior of a particle moving under the influence of a potential-energy function that is capable of binding it only if its total energy is below a certain value. The harmonic-oscillator potential does not have this property since it can bind a particle of arbitrarily high energy, but the potential $V(x)$ shown in Fig. 8-7 does. It is called a *finite square-well potential* and is used as a one-dimensional approximation to the potential function for an electron moving through a piece of metal or a neutron moving through a nucleus. If the total energy E of the particle is lower than the rim of the potential at V_0, the particle will be bound. If E is higher than V_0, it will not be bound by the potential; all regions of the x axis will be accessible to it in classical mechanics since $K = [E - V(x)] > 0$ for all x if $E > V_0$.

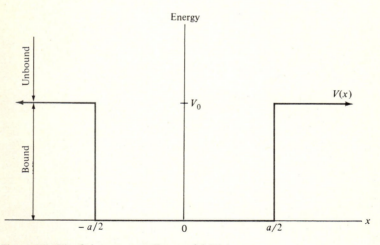

Figure 8-7. The finite square-well potential $V(x)$.

The potential can be expressed as

$$V(u) = \begin{cases} V_0 & u < -\dfrac{1}{2} \quad \text{or } u > \dfrac{1}{2} \\[2mm] \dfrac{V_0}{2} & u = \pm\dfrac{1}{2} \\[2mm] 0 & -\dfrac{1}{2} < u < \dfrac{1}{2} \end{cases} \qquad (8\text{-}20)$$

where the convenient dimensionless variable

$$u = \frac{x}{a} \qquad (8\text{-}21)$$

has been introduced. With this you have

$$\frac{d^2\psi}{dx^2} = \frac{d^2\psi}{d(au)^2} = \frac{1}{a^2}\frac{d^2\psi}{du^2}$$

and the Schroedinger equation for the potential is, from Eq. (8-8)

$$\frac{d^2\psi}{du^2} = -\frac{8\pi^2 ma^2}{h^2}[E - V(u)]$$

or $$\frac{d^2\psi}{du^2} = \begin{cases} -\beta(\epsilon - 1)\psi & u < -\dfrac{1}{2} \quad \text{or } u > \dfrac{1}{2} \\[2mm] -\beta\left(\epsilon - \dfrac{1}{2}\right)\psi & u = \pm\dfrac{1}{2} \\[2mm] -\beta\epsilon\psi & -\dfrac{1}{2} < u < \dfrac{1}{2} \end{cases} \qquad (8\text{-}22)$$

where $$\beta = \frac{8\pi^2 ma^2 V_0}{h^2} \qquad (8\text{-}23)$$

and $$\epsilon = \frac{E}{V_0} \qquad (8\text{-}24)$$

The numerical solution of the differential equation (8-22) is programmed just as it is for the harmonic oscillator, except: (1) there are the necessary changes to give the coefficient C in the general expression $d^2\psi/du^2 = C$ the forms defined in (8-22); and (2) there is no provision for skipping the display, as this feature will not be needed and would be difficult to provide because so much programming capacity is used for the routings required to generate the various forms for C.

The programs are listed in Tables 8-2 (SR-56) or (HP-25). They are used in the same way as the preceding ones, except: (1) the parameter β is preloaded in register 4 instead of the display skip index j; and (2) the value of Δu must be such that it divides evenly into 0.5 because the programs assume that one of the values of u_i will be precisely equal to 0.5.

TABLE 8-2. (SR-56) Finite square-well Schroedinger equation

Register Contents:

0	1	2	3	4	5	6	7	8	9
$\beta\epsilon$	ψ	$d\psi/du$	Δu	β	$\beta(\epsilon - 1/2)$	$\beta(\epsilon - 1)$	u	$C, C/2$	—

preloaded
start with ϵ in display register;
Δu must divide evenly into 0.5

Program

Step	Code	Key Entry	Comments
00	64	x	
01	34	RCL	
02	04	4	β
03	94	=	$\epsilon\beta$
04	33	STO	
05	00	0	0 now $\beta\epsilon$
06	74	−	
07	34	RCL	
08	04	4	$(\epsilon - 1)\beta$
09	94	=	$(\epsilon - 1)\beta$
10	33	STO	
11	06	6	6 now $\beta(\epsilon - 1)$
12	84	+	
13	34	RCL	β
14	04	4	β
15	54	÷	
16	02	2	2
17	94	=	$(\epsilon - 1/2)\beta$
18	33	STO	
19	05	5	5 now $\beta(\epsilon - 1/2)$
20	00	0	

Step	Code	Key Entry	Comments
49	03	3	Δu
50	64	x	
51	34	RCL	
52	08	8	$C/2$ in prelim. loop, C in next loop
53	94	=	$\Delta u C/2$ in prelim. loop, $\Delta u C$ in next loop
54	35	SUM	
55	02	2	2 now $d\psi/du_{1/2}$ in prelim. loop, $d\psi/du_{3/2}$ in next
56	34	RCL	
57	03	3	Δu
58	35	SUM	
59	07	7	7 now u_1 in prelim. loop, u_2 in next
60	64	x	
61	34	RCL	
62	02	2	$d\psi/du_{1/2}$ in prelim. loop, $d\psi/du_{3/2}$ in next
63	94	=	$\Delta u\, d\psi/du_{1/2}$ in prelim. loop, $\Delta u\, d\psi/du_{3/2}$ in next

Step	Code	Key	Comment
21	33	STO	
22	07	7	7 zeroed
23	34	RCL	Start loop
24	07	7	u
25	32	x ≤ t	Test register loaded with u
26	02	2	2
27	20	2nd 1/x	1/2
28	12	INV	
29	47	2nd x ≥ t	Test if $u > 1/2$
30	07	7	
31	03	3	To obtain $\beta(\epsilon - 1)$
32	37	2nd x = t	Test if $u = 1/2$
33	07	7	
34	08	8	To obtain $\beta(\epsilon - 1/2)$
35	34	RCL	
36	00	0	$\beta\epsilon$
37	64	×	
38	34	RCL	
39	01	1	ψ
40	94	=	$\beta\epsilon\psi$, or $\beta(\epsilon - 1)\psi$, or $\beta(\epsilon - 1/2)\psi$
41	93	+/-	C
42	33	STO	
43	08	8	8 now C
44	00	0	0
45	37	2nd x = t	Test u for routing in preliminary loop
46	08	8	
47	03	3	
48	34	RCL	
64	35	SUM	1 now ψ_1 in prelim. loop, ψ_2 in next
65	01	1	
66	34	RCL	ψ_1 in prelim. loop, ψ_2 in next
67	01	1	
68	59 (or 41)	2nd PAUSE (or R/S)	
69	59 (or 46)	2nd PAUSE (or 2nd NOP)	
70	22	GTO	To new loop
71	02	2	
72	03	3	
73	34	RCL	$\beta(\epsilon - 1)$
74	06	6	
75	22	GTO	To continue loop
76	03	3	
77	07	7	
78	34	RCL	$\beta(\epsilon - 1/2)$
79	05	5	
80	22	GTO	To continue loop
81	03	3	
82	07	7	
83	02	2	2
84	12	INV	
85	30	2nd PROD	
86	08	8	8 now $C/2$
87	22	GTO	
88	04	4	
89	08	8	To continue loop

Table 8-2. (HP-25) Finite square-well Schroedinger equation

Register Contents:

0	1	2	3	4	5	6	7
$\beta\epsilon$	ψ	$d\psi/du$	Δu	β	$\beta(\epsilon - 1/2)$	$\beta(\epsilon - 1)$	u

preloaded
start with ϵ in X register;
Δu must divide evenly into 0.5

Program

Step	Code	Key Entry	X	Y	Z	T	Comments
00			ϵ				
01	24 04	RCL 4	β	ϵ			
02	61	x	$\epsilon\beta$				
03	23 00	STO 0	$\epsilon\beta$				0 now $\beta\epsilon$
04	24 04	RCL 4	β	$\epsilon\beta$			
05	41	−	$(\epsilon - 1)\beta$				
06	23 06	STO 6	$(\epsilon - 1)\beta$				6 now $\beta(\epsilon - 1)$
07	24 04	RCL 4	β	$(\epsilon - 1)\beta$			
08	02	2	2	β	$(\epsilon - 1)\beta$		
09	71	÷	$\beta/2$	$(\epsilon - 1)\beta$			
10	51	+	$(\epsilon - 1/2)\beta$				
11	23 05	STO 5	$(\epsilon - 1/2)\beta$				5 now $\beta(\epsilon - 1/2)$
12	00	0	0				
13	23 07	STO 7	0				7 zeroed
14	24 07	RCL 7	u	u			Start loop
15	02	2	2	u			
16	15 22	g 1/x	1/2	u			
17	14 41	f x < y	1/2	u			Test u for routing to obtain $\beta(\epsilon - 1)$
18	13 39	GTO 39	1/2	u			
19	14 71	f x = y	1/2	u			Test u for routing to obtain $\beta(\epsilon - 1/2)$
20	13 41	GTO 41	1/2	u			

164

Line	Code	Keys	(Y)	(X)	Comment
21	24 00	RCL 0	$\beta\epsilon$	$\beta\epsilon$	Will be $\beta(\epsilon-1)$ or $\beta(\epsilon-1/2)$ if appropriate
22	24 01	RCL 1		ψ	
23	61	x		$-C$	$C = -\beta\epsilon\psi$, or $-\beta(\epsilon-1)\psi$, or $-\beta(\epsilon-1/2)\psi$
24	24 07	RCL 7	$-C$	u	Test u for routing to put 1/2 in
25	15 71	g x = 0	$-C$	u	preliminary loop
26	13 43	GTO 43		u	
27	22	R↓		$-C$	Will be $-C/2$ in prelim. loop
28	24 03	RCL 3	$-C$	Δu	2 now $d\psi/du_{1/2}$ in prelim. loop
29	61	x		$-C\Delta u$	2 now $d\psi/du_{3/2}$ in next loop
30	23 41 02	STO – 2	$-C$	$-C\Delta u$	
31	24 03	RCL 3		Δu	7 now u_1 in prelim. loop
32	23 51 07	STO + 7		Δu	7 now u_2 in next loop
33	24 02	RCL 2	Δu	$d\psi/du$	1 now ψ_1 in prelim. loop
34	61	x		$\Delta u\, d\psi/du$	1 now ψ_2 in next loop
35	23 51 01	STO + 1		$\Delta u\, d\psi/du$	
36	24 01	RCL 1		ψ	
37	14 74 (or 74)	f PAUSE (or R/S)		ψ	
38	13 14	GTO 14		ψ	To new loop
39	24 06	RCL 6		$\beta(\epsilon-1)$	To continue loop
40	13 22	GTO 22		$\beta(\epsilon-1)$	
41	24 05	RCL 5		$\beta(\epsilon-1/2)$	To continue loop
42	13 22	GTO 22		$\beta(\epsilon-1/2)$	
43	22	R↓		$-C$	
44	02	2		2	
45	71	÷	$-C$	$-C/2$	
46	13 28	GTO 28		$-C/2$	To continue loop

Examples

initial $\psi = 1$, initial $d\psi/du = 0$, $\Delta u = 0.05$, $\beta = 64$, $\epsilon = 0.0986$

initial $\psi = 1$, initial $d\psi/du = 0$, $\Delta u = 0.05$, $\beta = 64$, $\epsilon = 0.0987$

The results obtained by running with these values can be seen in Fig. 8-8. They define the shape of the wavefunction for the first energy level of a square well whose strength parameter is $\beta = 8\pi^2 ma^2 V_0/h^2 = 64$. Note that this wavefunction is even in u, just as was the wavefunction for the first energy level of an oscillator. The general shapes are similar for the two potentials, but the divergence to positive or negative infinity is less pronounced for the square well. Why? The values of ϵ show that the lowest energy level of this square well is at the energy

$$E \simeq 0.0986 \ V_0$$

where Eq. (8-24) has been used. ////

Examples

initial $\psi = 0$, initial $d\psi/du = 5.00$, $\Delta u = 0.02$, $\beta = 64$, $\epsilon = 0.382$

initial $\psi = 0$, initial $d\psi/du = 5.00$, $\Delta u = 0.02$, $\beta = 64$, $\epsilon = 0.383$

The initial value of $d\psi/du$ for this odd wavefunction was chosen so that its peak value when plotted in Fig. 8-9 is approximately equal to the peak value of the wavefunction plotted in Fig. 8-8. This example finds the wavefunction and allowed energy for the second energy level of the $\beta = 64$ square well. The energy level is at

$$E \simeq 0.382 \ V_0 \qquad \text{////}$$

You should search for discrete levels of this square well at higher energy. If you do, you will succeed in finding a third of the same general character as the first and second levels, except that its wavefunction has more oscillations within the confines of the well. But you will not succeed in finding a fourth discrete energy level for the $\beta = 64$ square well. The well does have a fourth level, in a certain sense that will be explained in the next section, but it occurs at too high an energy for it to be bound by the well. That is, it is found at a value of E above the rim of the potential well.

8-5. CONTINUUM SOLUTIONS AND VIRTUAL LEVELS

The next example gives you a solution to the finite-square-well Schroedinger equation for $\epsilon = E/V_0$ chosen to have a typical value somewhat greater than 1. That is, E is greater than V_0, so in classical mechanics the particle would be unbound. (See Fig. 8-7.)

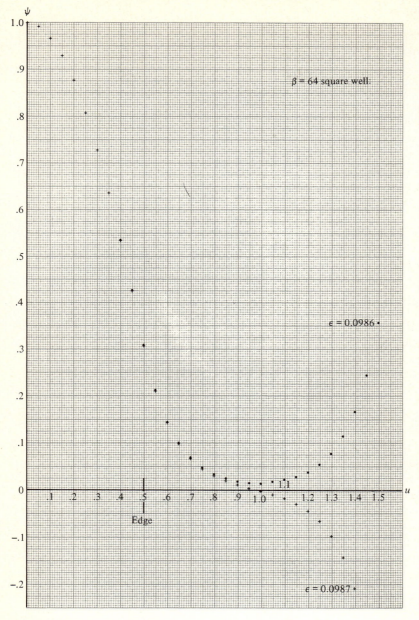

Figure 8-8. The first wavefunction of a $\beta = 64$ square well.

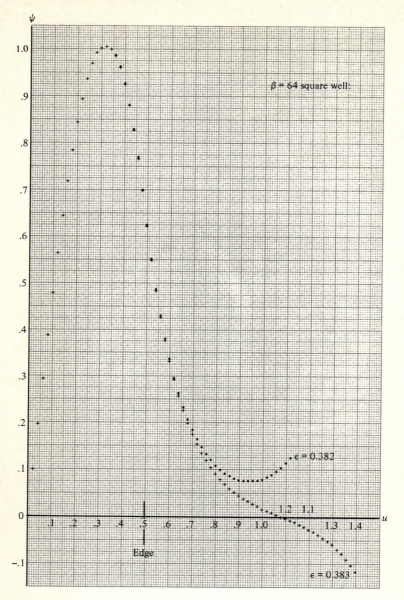

Figure 8-9. The second wavefunction of a $\beta = 64$ square well.

Example

initial $\psi = 1$, initial $d\psi/du = 0$, $\Delta u = 0.02$, $\beta = 64$, $\epsilon = 1.4$

Figure 8-10 shows that the solution ψ is oscillatory in the region within the well. The reason is the same as the reason why the differential equation for a classical harmonic oscillator has an oscillatory solution. In fact, except for the different independent and dependent variables, it is essentially the same differential equation:

$$d^2\psi/du^2 = -\beta\epsilon\psi$$

where $\beta\epsilon$ is a positive constant. So, within the well ψ is a sinusoidal. The same is true of ψ in the region outside the well, and of the differential equation it satisfies there:

$$d^2\psi/du^2 = -\beta(\epsilon - 1)\psi$$

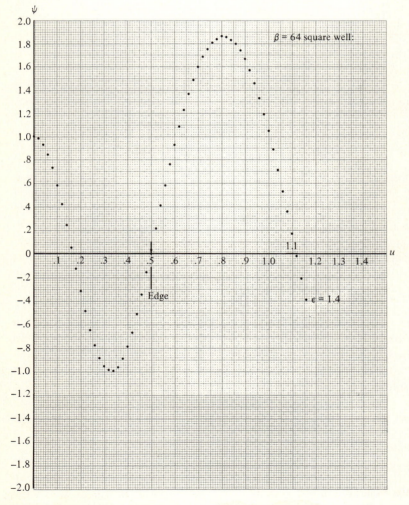

Figure 8-10. Typical even continuum wavefunction of a $\beta = 64$ square well.

since $\beta(\epsilon - 1)$ is another positive constant. The exterior part of ψ oscillates less rapidly than the interior part because $\beta(\epsilon - 1)$ is smaller than $\beta\epsilon$. As a consequence, the total ψ consists of a more rapidly oscillating sinusoidal joined smoothly at the edge of the well on to a less rapidly oscillating sinusoidal. You should note that for the typical case illustrated in Fig. 8-10, the amplitude of the exterior oscillation is larger than that of the interior oscillation.   ////

It is even more important for you to note that there is no difficulty at all in obtaining a solution to the Schroedinger equation for an energy parameter $\epsilon > 1$ (corresponding to an energy $E > V_0$) which is acceptable from the point of view of Born's postulate. The solution remains oscillatory no matter how large the u, and no matter what the particular value of E, so there is no tendency for ψ to diverge to infinity when the energy E of the particle is above the rim of the well. Since classically such a particle would not be bound within the well, you can conclude that: When the relation between the potential energy V of a particle and its total energy E is such that in classical mechanics the particle would not be bound to a limited region of space, then in quantum mechanics all values of E are allowed. These allowed values of E are said to form a *continuum*.

Example

initial $\psi = 1$, initial $d\psi/du = 0$, $\Delta u = 0.02$, $\beta = 64$, $\epsilon = 2.5$

All values of E or ϵ in the continuum above V_0 or 1 are equal in the sense that all of them are allowed. But Fig. 8-11, which displays the ψ obtained in this example, shows that some values of energy in the continuum are more equal than others! For these special energies the curvature of the interior part of ψ has such a value that $d\psi/du = 0$ at the edge of the well. The result is that the amplitude of the exterior oscillation is the same as the amplitude of the interior oscillation. To be sure you appreciate the point, compare Fig. 8-11 with Fig. 8-10.   ////

Both of these figures show continuum wavefunctions which are even functions of u. There are also continuum wavefunctions which are odd functions of u. In fact, at any energy E in the continuum starting at $E = V_0$, there are two allowed wavefunctions, one being even and the other odd.

Example

initial $\psi = 0$, initial $d\psi/du = 9.45$, $\Delta u = 0.02$, $\beta = 64$, $\epsilon = 1.4$

Figure 8-12 shows the odd wavefunction obtained from this example. It is for the same value of the energy parameter ϵ as was used for the even wavefunction shown in Fig. 8-10. To make comparison easier, the initial value of $d\psi/du$ was adjusted so that the amplitude of the interior oscilla-

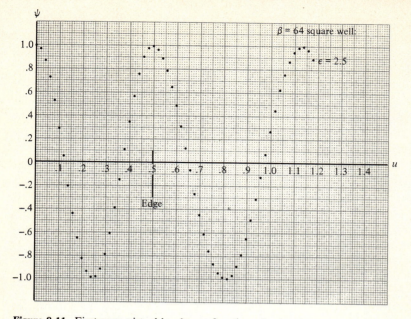

Figure 8-11. First even virtual-level wavefunction of a $\beta = 64$ square well.

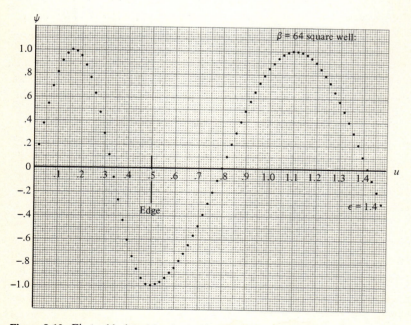

Figure 8-12. First odd virtual-level wavefunction of a $\beta = 64$ square well.

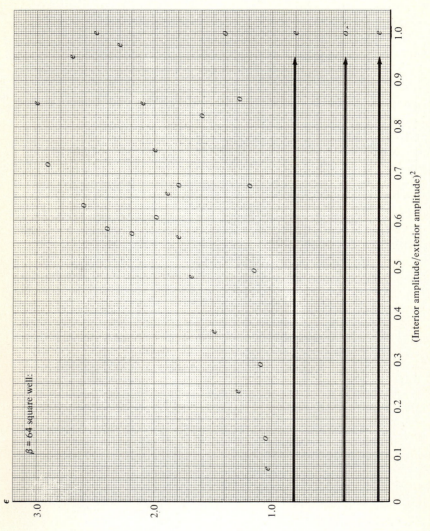

Figure 8-13. The three bound levels and the first two virtual levels of a $\beta = 64$ square well.

172

tion here is the same as it is for the two preceding examples. The point to note is that because the value of ϵ leads to a curvature which makes the condition $d\psi/du = 0$ satisfied at $u = \frac{1}{2}$, the exterior- and interior-oscillation amplitudes are equal. The values of ϵ producing this condition are not the same for the even and odd wavefunctions, as the two cases have quite different initial conditions at $u = 0$. ////

The reason why the ratio of the interior to exterior amplitudes is interesting is that Born's postulate says its square is related to the ratio of the probability of finding the particle within the well to the probability of finding it, in a range of u of equal length, outside the well. Figure 8-13 displays values of the square of the ratio of the amplitudes found in the preceding examples, plus values obtained by running the program with both even and odd initial conditions for a number of other choices of ϵ in the range 1.0 to 3.0. Each of the symbols e (for even) or o (for odd) are plotted so that its location gives the value of the ratio squared on the horizontal scale and the value of the energy parameter on the vertical scale.

Also shown on Fig. 8-13 are the three bound energy levels of this $\beta = 64$ square well. They are represented, in the usual fashion, by horizontal lines located at the proper values of ϵ and are labeled e or o to show whether they are even or odd in u. The arrow heads are to remind you that, although for bound levels an exterior amplitude cannot be defined, for the wavefunctions of such levels a number measuring the ratio of the probability of finding the particle inside the well to the probability of finding it outside would have a large value.

The figure gives you an overall picture of the bound- and lower-lying continuum states of this particular square well. It makes it clear that, although any energy is allowed in the continuum, the continuum is not structureless. In fact, the fairly sharp peak in the ratio for odd wavefunctions at $\epsilon \simeq 1.4$, and the somewhat broader peak for even wavefunctions at $\epsilon \simeq 2.5$, are sometimes called *virtual levels*. You can also see from the figure that there are obvious relations between the bound and virtual levels of the potential well.

If you would like to study a detailed development related to the subject of this chapter or read about other aspects of the fascinating field of quantum mechanics, Ref. 4 is recommended.

EXERCISES

8-1. Run the program and find the third wavefunction and allowed energy value of the harmonic oscillator.

8-2. Run the program and find the third bound wavefunction and energy value of the $\beta = 64$ square well.

8-3. Show that there is no fourth bound wavefunction of the $\beta = 64$ square well.

8-4. By trying larger values of β, find a value at which the fourth wavefunction of the square well becomes bound.

8-5. By running the program for several successively smaller values of β, show that no matter how small it is, i.e., no matter how weak the well, there is always one bound wavefunction for a one-dimensional square well. Explain why.

8-6. Find the second even and second odd virtual levels of a $\beta = 64$ square well. In the process, extend the plot of (interior amplitude/exterior amplitude)^2 shown in Fig. 8-13.

8-7. Using a large value of β, say $\beta = 1000$, find the first three wavefunctions and energy values of an approximation to an infinitely deep square well. Plot the wavefunctions and make a comparison of them and of their energy values with the corresponding ones of the $\beta = 64$ square well. Consider particularly the behavior of the wavefunctions in the region outside the well. Does this behavior suggest a simpler way to treat the problem of the infinitely deep square well?

8-8. Study the details of the programs and explain precisely what happens in each step.

8-9. Modify the harmonic-oscillator program to display ψ^2 by adding an x^2 instruction before the display instruction. Then use it to plot ψ^2 for the first, third, and ninth wavefunctions. Now go back to the classical oscillator program in Chap. 4 and modify it to display $(dx/dt)^{-1}$, and plot. To avoid problems with calculating $\frac{1}{0}$, use initial conditions $x = 0$, $dx/dt = 1$. When the oscillator comes to the end of its swing, the program will stop for you automatically if you get a $\frac{1}{0}$. Otherwise, stop it manually. Compare $(dx/dt)^{-1}$ with ψ^2 and discuss the relation between the behavior of the classical and quantum-mechanical oscillators for higher and higher energy values of the latter.

8-10. Follow the suggestion mentioned in 8-7 to formulate in a simple way the problem of the infinitely deep square well; then write a program to solve the Schroedinger equation for it and use the program to find the first three wavefunctions and energy values. Plot the wavefunctions and compare them and the energy values with 8-7.

8-11. Modify the program of 8-10 to display ψ^2, as in 8-9. Explain why $(dx/dt)^{-1}$ for the corresponding classical problem would be expected to be constant within the well in this case, and then compare it with ψ^2.

8-12. Write a program to solve the Schroedinger equation for an anharmonic-oscillator potential

$$V(x) = \frac{kx^2}{2} + \frac{px^4}{4}$$

First reexpress the Schroedinger equation in the dimensionless form

$$\frac{d^2\psi}{dx^2} = -(\epsilon - u^2 - \delta\, u^4)\psi$$

and find the relation between δ and p. Incorporate in your program provision for displaying ψ^2. Then find the ninth wavefunction and energy value and compare the latter with that found for the harmonic oscillator in the example. Plot ψ^2, compare the result with 8-9, and explain the difference. There is no analytical solution to this Schroedinger equation.

8-13. The Saxon-Woods potential, in one dimension and with distance expressed in units of the well width a, is

$$V(u) = \frac{-V_0}{1 + e^{(|u|-0.5)/\delta}}$$

It is used in nuclear physics to represent, more accurately than a square well, the potential acting between a nucleus and a passing neutron. Write a small program to plot $V(u)$ for the typical value $\delta = 0.05$ and compare it with the square-well potential.

8-14. Write a program to solve the Schroedinger equation for the Saxon-Woods potential of 8-13. Find the three bound wavefunctions and energy values for $8\pi^2 ma^2 V_0/h^2 = 64$ and compare them with the corresponding ones for the square well of the same strength. Explain the small differences. There is no analytical solution to this Schroedinger equation.

REFERENCES

1. Bueche, Frederick J.: *Introduction to Physics for Scientists and Engineers,* 2d ed., McGraw-Hill Book Company, New York, 1975, pp. 723–724.

   Eisberg, Robert, and Robert Resnick: *Quantum Physics of Atoms, Molecules, Solids, Nuclei, and Particles,* John Wiley & Sons, Inc., New York, 1974, p. 63.

   Halliday, David, and Robert Resnick: *Fundamentals of Physics,* John Wiley & Sons, Inc., New York, 1974, p. 781.

2. Bueche, Frederick J.: *Introduction to Physics for Scientists and Engineers,* 2d ed., McGraw-Hill Book Company, New York, 1975, p. 725.

   Eisberg, Robert, and Robert Resnick: *Quantum Physics of Atoms, Molecules, Solids, Nuclei, and Particles,* John Wiley & Sons, Inc., New York, 1974, p. 64.

   Halliday, David, and Robert Resnick: *Fundamentals of Physics,* John Wiley & Sons, Inc., New York, 1974, p. 782.

3. Eisberg, Robert, and Robert Resnick: *Quantum Physics of Atoms,*

Molecules, Solids, Nuclei, and Particles, John Wiley & Sons, Inc., New York, 1974, p. 72.

Halliday, David, and Robert Resnick: *Fundamentals of Physics,* John Wiley & Sons, Inc., New York, 1974, p. 789.

4. Eisberg, Robert, and Robert Resnick: *Quantum Physics of Atoms, Molecules, Solids, Nuclei, and Particles,* John Wiley & Sons, Inc., New York, 1974.